High-Temperature Superconducting Materials

High-Temperature Superconducting Materials

PREPARATIONS, PROPERTIES, AND PROCESSING

edited by

William E. Hatfield
John H. Miller, Jr.

University of North Carolina
Chapel Hill, North Carolina

Marcel Dekker, Inc. • New York and Basel

0298-1683

PHYSICS

Library of Congress Cataloging in Publication Data

High-temperature superconducting materials.

   Includes index.
   1. Superconductors.   I. Hatfield, William E.
II. Miller, John H.
TK7872.S8H54   1988   620.1'12973   88-3746
ISBN 0-8247-7995-9

MARCEL DEKKER, INC.
270 Madison Avenue, New York, New York 10016

Current printing (last digit):
10 9 8 7 6 5 4 3 2 1

PRINTED IN THE UNITED STATES OF AMERICA

Foreword

One of the most exciting developments in science in recent times is the discovery during the past year of high-temperature superconducting oxides. It has captured the imagination of the public with front-page stories in newspapers such as The New York Times and The Wall Street Journal and cover stories in magazines such as Time and Newsweek. A superconductor carries a current without resistance. A current flowing around a superconducting ring will flow indefinitely as long as the temperature is maintained below that for the transition from the normal to the superconducting state.

Superconductors are now used for a wide variety of applications, from very sensitive detectors of magnetic fields and electromagnetic radiation to high-field superconducting magnets for magnetic resonance imaging tomography and particle accelerators. The largest installation in the world is at the Fermi National Accelerator Laboratory west of Chicago where

---

Excerpt from the Van Vleck Memorial Lecture given at the University of Minnesota, October 1987.

there is a four-mile ring of superconducting magnetics. Elementary particles, protons, accelerated to very high energies, are confined to flow around the ring by the magnetic fields produced by the magnets. Applications have been limited by the difficulty and cost of cooling to helium temperatures.

Superconductivity was discovered in 1911, a little over seventy-five years ago, by Kamerlingh Onnes, at the University of Leiden in Holland. He found that the resistance of a rod of frozen mercury dropped to zero when cooled to the temperature of liquid He. It was soon found that many other elements, alloys and intermetallic compounds become superconducting when cooled to sufficiently low temperatures. Over the years, the highest transition temperature had been gradually increased from the 4 K of Hg to 23 K in the compound Nb_3Ge, discovered by scientists at Westinghouse about 15 years ago.

This remained the record until about a year ago, when two scientists, K. A. Müller and J. G. Bednorz, working at the IBM laboratory in Zurich, reported possible superconductivity in a mixture of La and Ba copper oxides at temperatures as high as 30 K, although true superconductivity, zero resistance, was found only below about 15 K. They received the 1987 Nobel Prize in Physics for their discovery.

Their work started the present explosion of interest, with compounds being discovered with transition temperatures in the range of 30-40 K, then another class in the range 90-100 K. There have been indications of superconductivity as high as 240 K, about $-10^{o}F$, a cold day in the winter in Minneapolis. Very recently there have been reports of possible room temperature superconductivity.

The new superconductors are ceramic oxides, not metals, and have the mechanical properties of ceramics.

They are brittle, not ductile like metals. Further,
the conduction is highly anisotropic. High current
densities are obtained only when the flow is along two-
dimensional sheets of copper oxide. The sheets are
separated by divalent and trivalent ions. Limiting
current densities are much lower when flow is
perpendicular to the sheets. High current densities
are not found in the usual polycrystalline material in
which the crystallites have random orientation. Real
innovation in materials technology will be required to
make the oxides practical for most potential
applications.

C. W. "Paul" Chu and associates at the University
of Houston began studying the new oxides in November,
1986. It was after a meeting at MIT in early December,
where both Chu and Japanese scientists reported on
their preliminary work, that physicists in this country
initiated the explosion of interest in this country and
the rest of the world. Chu's group not only isolated
the structure of a class of compounds with superconducting transitions around 30-40 K, but with an asso-
ciate and former student, M. K. Wu, at the University
of Alabama, discovered in January, 1987, a new class of
compounds with transition temperatures in the range 90-
100 K.

Prior to the early 1960s, superconductivity was
of interest only to scientists as a remarkable manifestation of quantum principles on a macroscopic scale.
Two important discoveries at that time led to two types
of applications called large scale and small scale.
The large-scale applications are those that make use of
the zero resistance. Intermetallic compounds like NbTi
and $NbSn_3$ were discovered that withstand very large
magnetic fields, making possible high-field superconducting magnets. These are used for a wide variety of
real or potential applications mentioned earlier, high-

energy physics, magnetic imaging, levitated trains,
magnetic energy storage, etc.

The second class of applications, called small
scale, is based on the quantum aspects such as flux
quantization. Superconducting computer circuits have
been built with loops so small that the presence or
absence of a flux quantum can be detected. In binary
logic, zero corresponds to no flux and unity to one
flux quantum. They would certainly be practical today
if it were not for the competition with semiconductor
chips in which up to a million transistors can be
placed on a tiny chip at low cost. Superconducting
elements are now used in by far the most sensitive
detectors for magnetic fields and various types of
electromagnetic radiation. Many of these small-scale
applications are based on the Josephson effect, which
depends on the wave aspects of electrons in superconductors. A supercurrent can flow through a thin
insulating layer separating two superconductors with a
magnitude that is dependent on the phase difference
between them. Phase is a variable associated with
waves. Josephson won the 1973 Nobel Prize for his
prediction of this effect before it was observed in the
laboratory.

Superconductivity is a macroscopic quantum
phenomenon. Quantum effects are exhibited on a
macroscopic scale rather than the scale of atoms and
molecules. There was no possibility of explaining the
phenomenon before the advent of quantum mechanics in
the mid-twenties. Quantum theory, in which matter has
both particle and wave aspects, discovered by Heisen-
berg and Schrödinger in 1924 and 1925, was soon applied
to transport of electricity by electrons in metals, but
superconductivity remained a mystery.

In 1950 it was shown that the transition temperature to the superconducting state depends on the
isotopic mass of atoms that make up the metal. This
suggested that superconductivity involves an interaction between the conduction electrons and the vibrational motion of the ions in the metal. Even with this
clue it took another seven years before a satisfactory
theory was developed by Leon Cooper, Robert Schrieffer
and me at the University of Illinois. It is generally
called the BCS theory after the initials of the
authors.

This theory was very successful not only in
explaining what was known about superconductivity but
in predicting new phenomena that later were confirmed
by laboratory experiments. The theory is based on a
coherent pairing of electrons such that all pairs have
identically the same momenta. The pairing results from
a long-range attraction between electrons.

In free space, electrons have the same charge and
repel one another. To have an attraction that gives
rise to the pairing, it is necessary to put the
electrons in an environment such that at large
distances there is an attractive force that dominates
the Coulomb repulsion. In a superconducting metal, the
force comes from motion of the ions that make up the
crystal lattice in which the electron moves. One can
think of one electron displacing a positive ion from
its equilibrium position. A second electron feels the
force of the displaced ion and is attracted to the same
vicinity as the first.

Over the years there have been many attempts to
find compounds that become superconducting at higher
temperatures. For seventy years, the maximum transition temperature was gradually increased from the 4 K

of Hg in 1911 to 23 K in Nb_3Ge in 1972, which remained
the record until the discovery of the new superconducting oxides during the past year. Metallic oxides were
first studied in the mid-seventies. Two groups of
oxides were found that had transition temperatures
around 13 K, remarkably high for the small concentration of electrons free to transport electricity. This
suggested to Müller and Bednorz that they try to
explore metallic oxides in the hope of finding high
transition temperatures.

Regardless of possible applications, the new
superconducting oxides are a remarkable scientific
discovery. There are about as many different expla-
nations for the observed high transition temperatures
as there are theorists who have thought about the
problem, and that is quite a large number. With more
complete experimental data on single crystals becoming
available, it should be possible to gradually weed
these out and arrive at the correct explanation.

The field of high-temperature superconductivity
is in a very dynamic research phase with many hundreds
of scientists throughout the world frantically search-
ing for the next breakthrough in performance or understanding. With so many competing groups there is
undoubtedly much duplication of effort, but with rapid
exchange of information through papers and conferences,
unnecessary duplication is kept to a minimum. Not
enough is known about what the future will bring to
have a more directed program. New unexpected discoveries occur frequently. It is a very exciting time
for those involved.

John Bardeen
Department of Physics
University of Illinois at Urbana-Champaign
Urbana, Illinois

Preface

The discovery of the onset of superconductivity
near 30 K in barium-doped lanthanum cuprate by Bednorz
and Müller immediately stimulated intense activity in
chemistry, physics, materials sciences, and engineering
laboratories around the world. One of the early key
advances in the research with these exciting new materials was the observation that T_c was dependent on
externally applied pressure. This led to the idea that
pressure could be applied internally by use of smaller
ions, in this case substitution of strontium for
barium, and the success of the experiment led to
additional ionic substitutions. Several new compounds
were produced, and many of these were polyphasic. A
major breakthrough was the discovery by Wu, Chu, and
coworkers of a material with T_c above 90 K. It is
remarkable that the 90 K superconductor was produced

less than three months after Bednorz and Müller's
discovery had been verified.

The advent of superconductors with transition
temperatures near 100 K and which have large critical
fields and critical currents will have a tremendous
impact on technology. Potential applications include
power transmission, levitated trains, motors and
generators, power storage, computers, and scientific
instruments, among others. Since it was clear that
this significant advance would have an immediate impact
on the scientific community, the North Carolina Section
of the American Chemical Society elected to organize a
multidisciplinary symposium devoted to preparations,
properties, and processing of high-temperature
superconducting materials. The symposium was preceded
by a tutorial in which important events that led to
high-temperature superconductors were presented in
chronological fashion, the basic physics of
superconductors especially as it relates to high-T_c
ceramic oxides was outlined, and the preparations,
properties, and electronic and crystal structures of
these new materials were described. These proceedings
consist of material from the tutorial lectures as well
as papers containing new results which were presented
at the symposium by invited lecturers and in
contributed poster presentations.

The symposium was organized by the North Carolina Section of the American Chemical Society and received support from the Section; Army Research Office, Durham; DuPont; AT&T Bell Laboratories; Westinghouse; IBM; the Curriculum in Applied Sciences at the University of North Carolina at Chapel Hill; and the Chemistry and Physics Departments of Duke University, North Carolina State University, and the University of North Carolina. Without this generous support, the symposium would not have been possible. The organizers wish to express their sincere appreciation to these organizations for their support and the opportunity to assemble many of the leaders in research on high-temperature superconducting materials for presentations of the most recent results on the preparation, properties, and processing of these exciting new materials.

The Organizing and Programming Committees consisted of Jerzy Bernholz (NCSU, Physics), William E. Estes (DuPont, Research Triangle Park), David Haase (NCSU, Physics), Herbert Hacker (Duke University, EE), Paul J. Kropp (UNC-CH, Chemistry and Curriculum in Applied Sciences), Valerie D. Kuck (AT&T Bell Laboratories), Robert Metzger (Alabama, Chemistry), Horst Meyer (Duke University, Physics), Richard A. Palmer (Duke University, Chemistry), Roger C. Sanwald (IBM, Research Triangle Park), Leonard W. ter Haar (AT&T Bell

Laboratories), M.-H. Whangbo (NCSU, Chemistry), and the
Editors of these proceedings. We wish to thank the
lecturers and participants for their contributions to
the symposium. Each manuscript was read by an
impartial reviewer. The authors and editors wish to
thank the referees for their advice.

William E. Hatfield

John H. Miller, Jr.

Contents

FOREWORD   iii
    John Bardeen

PREFACE   ix

1. OXIDE SUPERCONDUCTORS: STRUCTURE-PROPERTY
    RELATIONSHIPS AND MECHANISM FOR
    HIGH T_C   1
    A. W. Sleight

2. THE ADVENT OF HIGH-TEMPERATURE
    SUPERCONDUCTING MATERIALS: CHRONOLOGY
    OF EVENTS AND HALLMARK DEVELOPMENTS . .   37
    William E. Hatfield

3. HIGH-TEMPERATURE SUPERCONDUCTION:
    AN OVERVIEW   67
    Wid J. Painter

4. THE PHYSICS OF HIGH-TEMPERATURE
    SUPERCONDUCTIVITY   79
    John H. Miller, Jr.

5. DENSITY FUNCTIONAL THEORY FOR HIDDEN HIGH-T_C
    SUPERCONDUCTIVITY   99
    Akitomo Tachibana

6. WHERE CAN NEW CLASSES OF HIGH-T_C
    SUPERCONDUCTING MATERIALS BE FOUND? . .   107
    Francis J. DiSalvo

7. PRECIPITATION OF SUPERCONDUCTOR PRECURSOR
        POWDERS   121
    Bruce C. Bunker, James A. Voight,
    Daniel H. Doughty, Diana L. Lamppa,
    and Kathleen M. Kimball

8. SYNTHESIS AND CHARACTERIZATION OF HIGH-T_C
        SUPERCONDUCTORS IN THE
        $YBa_2(Cu_{1-x}Ni_x)_3O_{7-\delta}$ SYSTEM   131
    Teng-Ming Chen, Joseph F. Bringley,
    Bruce A. Averill, K. M. Wong, and
    S. Joseph Poon

9. PREPARATION AND PHYSICAL PROPERTIES OF
        OXYFLUORIDES IN THE Ln–Ba–Cu
        SYSTEM (Ln = Y, La)   141
    Joseph F. Bringley, Teng-Ming Chen,
    Bruce A. Averill, K. M. Wong, and
    S. Joseph Poon

10. OXALATE PRECIPITATION METHODS FOR PREPARING
        THE YTTRIUM-BARIUM-COPPER
        SUPERCONDUCTING COMPOUND   153
    Ronald J. Clark, William J. Wallace,
    and Jennifer A. Leupin

11. HIGH-TEMPERATURE OXIDE SUPERCONDUCTOR THICK
        FILMS   159
    Yonhua Tzeng

12. IODOMETRIC TITRATIONS AND XANES: TWO
        PERSPECTIVES OF THE COPPER VALENCE IN
        HIGH-T_C SUPERCONDUCTING OXIDES   167
    L. Soderholm, E. E. Alp, M. A. Beno,
    L. R. Morss, G. Shenoy, and G. L. Goodman

13. IMPORTANCE OF THE INTERACTIONS BETWEEN THE
        COPPER ATOMS OF THE CuO_2 LAYERS
        OCCURRING VIA THE OXYGEN–COPPER–OXYGEN
        ATOM BRIDGES OF THE CuO_3 CHAINS FOR
        THE HIGH–TEMPERATURE (T_C > 90 K)
        SUPERCONDUCTIVITY OF $YBa_2Cu_3O_{7-y}$ 181
    Myung–Hwan Whangbo, Michel Evain,
    Mark A. Beno, and Jack M. Williams

14. CRYSTAL STRUCTURE AND LATTICE VIBRATIONS OF
    THE CERAMIC SUPERCONDUCTOR
    $La_{1.85}Sr_{0.15}CuO_4$: NEUTRON SCATTERING
    STUDIES   211
    P. Day

15. OBSERVATION OF THE AC JOSEPHSON EFFECT INSIDE
    COPPER OXIDE-BASED SUPERCONDUCTORS . . .   235
    M. H. Devoret, D. Esteve, J. Martinis,
    C. Urbina, G. Collin, P. Monod,
    M. Ribault, and A. Revcolevschi

16. TEMPERATURE-DEPENDENT CONDUCTIVITY OF
    OXYGEN-DEPLETED YBCO CERAMICS   243
    J. H. Miller, Jr., B. Liu, W. J. Riley,
    A. N. Dibianca, S. L. Holder, J. D. Dunn,
    B. R. Rohrs, and W. E. Hatfield

17. EFFECT OF SURFACE PROPERTIES ON METAL-OXIDE
    SUPERCONDUCTOR CONTACT RESISTANCE . . .   251
    Yonhua Tzeng

18. OBSERVATIONS OF THE DETERIORATION OF $YBa_2Cu_3O_7$
    USING A NEW SUPERCONDUCTOR
    CHARACTERIZATION CRYOSTAT   261
    Ralph C. Longsworth and William A. Steyert

19. SUPERCONDUCTIVITY, THERMOGRAVIMETRY, EPR,
    ELECTRON MICROSCOPY, AND X-RAY
    DIFFRACTION OF $YBa_2Cu_3O_{7-z}$ 267
    Anny Morrobel-Sosa, David A. Robinson,
    Chinnarong Asavaroengchai,
    Robert M. Metzger, Joseph S. Thrasher,
    Chester Alexander, Jr., Donald A. Stanley,
    and M. Abbot Maginnis

20. OPTICAL PROPERTIES OF $La_{2-x}Sr_xCuO_4$ 275
    S. L. Herr, K. Kamarás, C. D. Porter,
    M. G. Doss, D. B. Tanner, D. A. Bonn,
    J. E. Greedan, C. V. Stager, T. Timusk,
    E. Etemad, D. E. Aspnes, M. K. Kelly,
    R. Thompson, J.-M. Tarascon, and
    G. W. Hull

21. INFRARED REFLECTANCE OF RARE EARTH-BARIUM-
    COPPER OXIDE SUPERCONDUCTORS   283
    R. Sudharsanan, S. Perkowitz, B. Lou,
    R. R. Caldwell, and G. L. Carr

22. SIMULATION OF CRYSTAL STRUCTURES BY EMPIRICAL
        ATOM-ATOM POTENTIALS.
        IV. INTERPRETATION OF THE RAMAN
        AND INFRARED SPECTRA OF $La_{2-x}M_xCuO_4$ . .   289
     Michel Evain, Myung-Hwan Whangbo,
     John R. Ferraro, and Jack M. Williams

23. INVESTIGATION OF Y-Ba-Cu-O SUPERCONDUCTING
        MATERIALS BY POSITRON ANNIHILATION
        LIFETIME SPECTROSCOPY   305
     A. J. Hill, F. H. Cocks, U. M. Goesele,
     P. L. Jones, T. Y. Tan, and A. I. Kingon

24. MAGNETIC ANOMALIES IN THE RARE EARTH OXIDE
        SUPERCONDUCTORS $GdBa_2Cu_3O_{7-x}$ AND
        $YbBa_2Cu_3O_{7-x}$ 313
     William E. Hatfield, Brian R. Rohrs,
     Martin L. Kirk, Jeffrey H. Helms,
     Hyekyeong Ro, and Eric J. Williamsen

25. THE LOW-TEMPERATURE SPECIFIC HEAT OF
        $YBa_2Cu_3O_{6+\delta}$ COMPOUNDS 321
     D. G. Haase, R. Velasquez, and
     A. I. Kingon

26. PROCESSING AND SUPERCONDUCTING PROPERTIES OF
        OF $GdBa_2Cu_3O_{7-z}$ 327
     Chinnarong Asavaroengchai,
     Robert M. Metzger, David A. Robinson,
     Anny Morrobel-Sosa, Joseph S. Thrasher,
     Chester Alexander, Jr., Donald A. Stanley,
     and M. Abbot Maginnis

27. PROCESSING AND MICROSTRUCTURES OF
        $YBa_2Cu_3O_{7-\sigma}$ 335
     Angus I. Kingon, Sopa Chevacharoenkul,
     Stane Pejovnik, Ricardo Velasquez,
     Richard L. Porter, Thomas M. Hare,
     Hayne Palmour III, and David G. Haase

28. RAMAN SPECTROSCOPIC CHARACTERIZATION OF
        DIFFERENT PHASES IN THE Y-Ba-Cu-O SYSTEM   349
     B. H. Loo, M. K. Wu, D. H. Burns,
     A. Ibrahim, C. Jenkins, T. Rolin,
     Y. G. Lee, D. O. Frazier and F. Adar

INDEX   367

1

Oxide Superconductors: Structure-Property Relationships and Mechanism for High T_c

A. W. SLEIGHT Central Research and Development Dept.,
E. I. du Pont de Nemours and Company, Experimental
Station, Building 356, Wilmington, Delaware 19898

1.1 INTRODUCTION

Although the phenomenon of superconductivity has been
known since 1911, oxide superconductors only came on the
scene about twenty years ago (Table 1.1). Initially
oxide superconductors did not cause much excitement
because their T_c's were lower than those found for
intermetallic compounds. However, we now know that some
oxide superconductors make up a remarkable class of
superconductors, holding the records for highest T_c and
highest critical field by wide margins. Furthermore, it
appears that a new mechanism for superconductivity is
required to account for the behavior of these amazing
materials.

1.2 STRUCTURE-PROPERTY RELATIONSHIPS

1.2.1 NbO and TiO

The only binary oxides which superconduct are NbO and
TiO (3). Both of these materials have NaCℓ related

1

Table 1.1.   Oxide Superconductors[+]

	Oxide	T_c	Date Reported
Metal-metal			
	NbO	1.2 K	1965
	TiO	~2 K	1965
Metal-oxygen π^*			
	$SrTiO_{3-x}$	0.7 K	1964
	A_xWO_3	6 K	1965
	A_xMoO_3	4 K	1966
	A_xReO_3	4 K	1969
	$LiTi_2O_4$	13 K	1974
Metal-oxygen σ^*			
	Ag_7O_8X	1 K	1966
	$Ba(Pb,Bi)O_3$	13 K	1975
	$(La,A)_2Cu_2O_4$	40 K	1986
	$RBa_2Cu_3O_7$	95 K	1987

[+]see text for references

structures, and they both have T_c's near 1 K. The
structures for both NbO and TiO are highly defective
relative to the NaCℓ structure; there are vacancies on
both the cation and anion sublattices. In the case of
TiO, these vacancies are disordered and there is a
significant departure from stoichiometry. For NbO, the
vacancies are well ordered and the coordination for both
Nb and O is square planar. The high defect
concentration in both TiO and NbO is due to a
competition between metal-metal bonding interactions and
metal-oxygen antibonding interactions of a π type. The
defects alleviate the antibonding interaction to some
extent and allow for better metal-metal bonding. TiO
and NbO are the only oxide superconductors in which the
conduction band is based largely on a direct metal-metal
interaction. The metal-metal distance in TiO and NbO is
just under 3 Å.

1.2.2 $SrTiO_{3-x}$

The first ternary oxide found to be superconducting was
based on $SrTiO_3$ (4). Thus, the field of oxide
superconductivity started with the perovskite structure,
and this structure type has remained the most important
one for oxide superconductors. Stoichiometric $SrTiO_3$ is
an insulator, but on doping to provide carriers in the
$3d$ band, it becomes conducting and, in certain cases,
superconducting. The highest T_c observed is about
0.7 K. Although we commonly refer to electron carriers
in $SrTiO_{3-x}$ as electrons in the $3d$ band, the Ti-Ti
distance is too great (nearly 4 Å) to support metallic
conductivity. We know that this conduction band could
not exist if it were not for strong covalent bonding
between titanium and oxygen. This band may be referred

to as a π^* band on the basis of the Ti-O bond that gives rise to the conduction band. In the case of the new copper oxide based superconductors, we have a σ^* conduction band instead of a π^* conduction band.

1.2.3 Tungsten Bronzes

Metallic oxides of the type A_xWO_3 derive their name from the appearance that some of them possess. The A cation is generally an alkali or an alkaline earth cation. The structures of the tungsten bronzes are all related to that of perovskite; however, there are cubic, tetragonal, and hexagonal tungsten bronzes. All three structures are made up of WO_6 octahedra which share corners to build up a three-dimensional network. The cubic tungsten bronze structure is identical to the perovskite structure with partial occupancy on the A cation site. For A_xWO_3, as with $SrTiO_{3-x}$, insulating properties are observed when x is zero. With increasing x, conductivity develops. However, it is only the tetragonal and hexagonal tungsten bronzes which become superconducting (5). For those tungsten bronzes that become superconducting, T_c actually decreases with increasing x. One could expect the opposite from classical BCS theory because increasing x might lead to an increased density of states at the Fermi level. Thus we have the suggestion that even these oxides exhibit a nonclassical mechanism for superconductivity.

1.2.4 Molybdenum Bronzes

Oxides of the type A_xMoO_3 normally possess structures different from those of the tungsten bronzes. Some of these are metallic, but none are superconducting.

However at high pressure, it is possible to prepare K_xMoO_3 which is isostructural with the tetragonal tungsten bronzes (6), and this phase is superconducting (T_c ~4.2 K). Again as with the cubic tungsten bronzes, cubic molybdenum bronzes are not superconducting.

1.2.5 Rhenium Bronzes

ReO_3 itself has the ideal cubic perovskite structure but with no A cation. It has good metallic properties, but it is not superconducting. A solid solution of the type $W_{1-x}Re_xO_3$ was prepared at high pressure (7), and superconductivity was not found for any value of x.

Rhenium bronzes of the type A_xReO_3 were also prepared at high pressure (6). Again the cubic bronzes were not superconducting, but hexagonal $K_{0.3}ReO_3$ was superconducting (T_c ~3.6 K). It should be noted that the number of conduction electrons per formula unit is x for A_xWO_3 and A_xMoO_3 phases whereas it is (x+1) for A_xReO_3. This much greater concentration of conduction electrons in the rhenium phases has no apparent effect on superconducting properties.

1.2.6 $LiTi_2O_4$

This compound has the spinel structure and actually has a range of composition, $Li_{1+x}Ti_{2-x}O_4$. The T_c in this system can reach about 13 K (8). The spinel structure is very different from the perovskite structure, but the basic building blocks are nonetheless TiO_6 octahedra which share corners to build up the three dimensional network. The density of states at the Fermi level for this material is that expected by classical BCS theory for a T_c of about 13 K. Thus, this oxide superconductor has not been considered to be one requiring a new mechanism to rationalize its superconducting behavior.

1.2.7 Ba(Pb,Bi)O$_3$

Here we are back to the perovskite structure with superconductivity which can also reach about 13 K (9). This remains the highest T$_c$ for any superconductor not containing a transition element. Ba(Pb,Bi)O$_3$ superconductors are regarded by many as the first example of a new type of superconductivity which is also exhibited in the recently discovered copper oxide based superconductors. Batlogg (10), for example, emphasizes that it is only Ba(Pb,Bi)O$_3$, (La,A)$_2$CuO$_4$ and RBa$_2$Cu$_3$O$_7$ superconductors which have abnormally high T$_c$'s considering their low density of states at the Fermi level.

A schematic phase diagram for the BaPb$_{1-x}$Bi$_x$O$_3$ system is shown in Figure 1.1. For x = 0, BaPbO$_3$, metallic properties are observed since the Pb 6s conduction band just overlaps the O 2p valence band (Figure 1.2). As expected from simple electronegativity considerations, Pb-O bonds have a high degree of covalency. Thus the 6s band is rich in oxygen character, and the 2p band is correspondingly rich in lead character. Superconductivity is observed for BaPbO$_3$ with a T$_c$ of about 0.5 K (11). It should be noted that low temperature superconductivity in BaPbO$_3$ was found several years after high temperature superconductivity was reported in the Ba(Pb,Bi)O$_3$ system.

For x = 1, BaBiO$_3$, semiconducting properties were found from 4.2 K to its decomposition temperature of about 900°C. On this basis, a Ba$_2$Bi^{III}Bi^{V}O$_6$ formulation

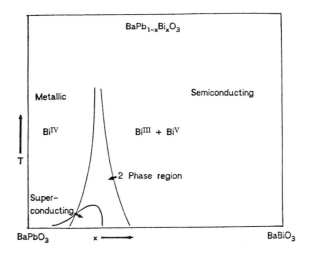

FIGURE 1.1 Schematic phase diagram for the $BaPb_{1-x}Bi_xO_3$ system. The two phase region is not normally observed.

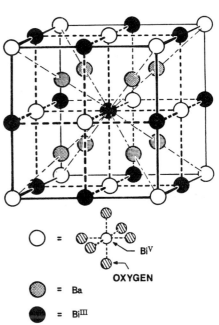

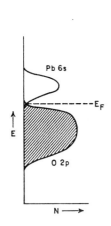

FIGURE 1.2 Schematic density of states *vs.* energy for $BaPbO_3$; E_F is the Fermi level.

FIGURE 1.3 The ordered perovskite structure of $Ba_2Bi^{III}Bi^VO_6$.

was considered likely, and this has been confirmed by
several studies (12,13). The Bi^{III} and Bi^V cations take
on an ordered arrangement in the perovskite structure
(Figure 1.3).

As x increases in the $BaPb_{1-x}Bi_xO_3$ system, the T_c
increases from 0.5 K at x = 0 to about 13 K at about x =
0.27. On further increasing x, superconductivity
rapidly disappears and the properties become those of a
semiconductor. In the metallic region, the oxidation
state of Bi is best described as tetravalent; the 6s
electron of Bi^{IV} is simply delocalized into a 6s band
made up of states from Bi(6s), Pb(6s) and O(2p).
Although all phases in the $BaPb_{1-x}Bi_xO_3$ system are
pseudocubic, the actual symmetry at room temperature is
lower than cubic for all values of x. Monoclinic
symmetry is found for $BaBiO_3$. Orthorhombic symmetry is
observed for $BaPbO_3$ and for the semiconducting region
from x ∼0.3 to x ∼0.9. The symmetry in the high T_c
region has been somewhat controversial, but recent
studies (14) have confirmed the tetragonal symmetry
originally indicated (15).

A two-phase region is shown in Figure 1.1, but this
does not mean that phases in this composition range are
necessarily mixtures of two phases with different x
values. In fact, the best superconductors are single
phases prepared by quenching from the single phase
region. Although such materials would become two phase
if annealed at some intermediate temperature, they
remain single phase at room temperature and below if
given an ordinary air quench. Thus we find that the
best superconductors in the $BaPb_{1-x}Bi_xO_3$ system are
metastable materials. Furthermore, we expect that even
in the single-phase high-T_c region there will be a
tendency for Pb and Bi segregation on a short range

basis. Presumably, this is the main reason why it is difficult to obtain sharp transitions in this system regardless of the value of x.

It is well known to chemists that the s^1 state, e.g. Bi^{IV}, Sb^{IV}, Sn^{III}, etc., is not a normal oxidation state. This state usually either leads to dimers (e.g. Sn^{III}-Sn^{III}) to pair the s electrons or to disproportionation into s^0 and s^2, also to pair the electrons. However, in dilute systems, s^1 cations can be observed and have been studied by ESR.

In the $BaPb_{1-x}Bi_xO_3$ system, the s^1 cation, Bi^{IV}, can exist when dilute on the Pb sites. However, as the Bi^{IV} concentration increases, the number of Bi-O-Bi linkages increases, and there is a strong drive to disproportionate into Bi^{III} (s^2) and Bi^V (s^0). When this happens, the conduction electrons are trapped by a local lattice distortion, and semiconducting properties develop. For x less than one, the Bi^{III} and Bi^V cations do not take on long range order because the Pb and Bi cations are disordered over the octahedral sites.

The first observation of superconductivity in the $BaPb_{1-x}Bi_xO_3$ system was made at Du Pont in 1974. The resistance of a sintered pellet of polycrystalline $BaPb_{0.75}Bi_{0.25}O_3$ was observed to fall to zero at about 12 K. The temperature dependence of the resistivity above T_c was that of a semiconductor, but we never published that result. We had previously seen this behavior (semiconductor-to-superconductor) in oxide bronzes. We felt that the semiconducting behavior was most likely arising from added resistance at grain boundaries. Thus we grew single crystals hydrothermally. These crystals showed superconducting properties, and above T_c the resistivity vs. temperature behavior was that of a metal (9). Still it remains a

question of considerable interest as to whether or not
any bulk superconductor shows semiconducting behavior
above T_c. There are reports (16) that $BaPb_{1-x}Bi_xO_3$
single crystals can be semiconducting above T_c.
However, this could be due to inhomogeneities in the
crystals with respect to the value of x. Further work
is needed in this area, especially since some samples of
the copper oxide based superconductors also show
semiconductivity above T_c. We need to understand
whether this is an intrinsic property of a bulk material
or whether it is due to inhomogeneities.

1.2.8 Ag_7O_8X Phases

Silver oxides of the type Ag_7O_8X with X = NO_3^-, HF_2^-,
ClO_4^- or BF_4^- are known to be superconducting (17). The
highest T_c of 1.4 K is found for X = HF_2^-. The
structure of these cubic Ag_7O_8X phases may be viewed as
a three dimensional network of square planar AgO_4 units
which share corners. There are Ag^{1+} and X^- ions placed
within the interstitial sites of this network. The
formula may then be written as $Ag^{1+}Ag_6O_8X^-$. The average
oxidation state of the Ag of the square planar network
is then +2.67. Thus, according to the mechanism
developed in section 1.3.2, these Ag_7O_8X phases should
be superconducting because square planar Ag^{II} has been
diluted with Ag^{III}.

1.2.9 $La_{2-x}A_xCuO_4$ Phases

The ideal tetragonal K_2NiF_4 structure is found for
La_2CuO_4 at high temperatures, but at room temperature
and below, there is a distortion to orthorhombic
symmetry (Figure 1.4). The infinite two dimensional
sheets containing square planar Cu^{II} are strictly planar

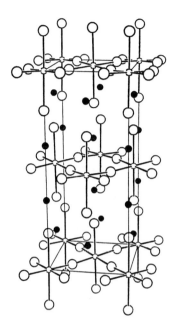

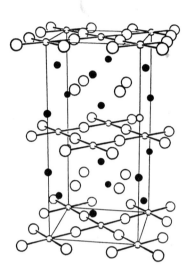

FIGURE 1.4   The structure
of La_2CuO_4.   The small open
spheres are Cu; the large
open spheres are O, and the
closed spheres are La.

FIGURE 1.5   The structure
of R_2CuO_4 phases where R
is Pr, Nd, Sm or Gd.   The
small open spheres are Cu;
the large open spheres are
O, and the closed spheres
are the R cations.

in the tetragonal structure but become slightly buckled
in the orthorhombic structure.   All copper atoms are
strictly equivalent to each other in both the tetragonal
and orthorhombic structures; thus, it is unfortunate
that this transition was proposed to be a Peierls
transition.   One does not need to do any calculations to
know that it cannot be a Peierls transition.   It is
enough to know that from a crystallographic point of
view, the number of unique atoms does not change during
the tetragonal-to-orthorhombic transition.

In fact, the tetragonal-to-orthorhombic transition
is a particularly easy one for chemists to appreciate.
In the tetragonal structure, the angle for the very
strong Cu-O-Cu bonds is exactly 180°. In the
orthorhombic structure, this angle deviates from 180°
for the same reason as in an ether such as CH_3OCH_3. A
180° π interaction is antibonding for both ethers and
Cu-O-Cu. This antibonding interaction can be
substantially decreased by bending the bonds.
Rehybridization occurs on the oxygen with the result
that the problematic p orbital on oxygen mixes with the
s orbital to form nonbonding lone pairs of electrons on
oxygen. Bending M-O-M bonds is the natural way to
alleviate π antibonding interactions. Thus, the
tetragonal-to-orthorhombic transition in La_2CuO_4 is no
mystery. The fact that this transition has practically
no effect on electrical properties is expected since the
π^* band is far below the Fermi level.

It should be noted that the tendency to bend
Cu-O-Cu bonds becomes stronger as the Cu-O distance
decreases. In the R_2CuO_4 phases where R is Pr, Nd, Sm
or Gd, the Cu-O-Cu angle remains strictly 180° because
the Cu-O distance is much longer (~1.98 Å) than in
La_2CuO_4 (~1.89 Å). As the Cu-O distance becomes longer,
the Cu-O orbital overlap decreases to the point where
the antibonding argument gives way to ionic
considerations which favor a 180° Cu-O-Cu bond.

The properties of La_2CuO_4 itself have been highly
controversial. The controversy is largely due to the
fact that the compound can be nonstoichiometric in
several different ways. Most of the attention has been
focused on the oxygen stoichiometry, i.e. La_2CuO_{4-x}.
However, there is also evidence for lanthanum
deficiency, i.e. $La_{2-y}CuO_{4-x}$. The lanthanum deficiency

in turn tends to give intergrowth phases (18) of the type $La_2CuO_4 \cdot nLaCuO_3$, i.e. a perovskite $LaCuO_3$ intergrowth with K_2NiF_4 type La_2CuO_4. Despite the confusion caused by this nonstoichiometry, there is currently a consensus that when La_2CuO_4 is very close to stoichiometry, it develops semiconducting properties at low temperatures. This fact leads to another problem since all band structure calculations indicate that La_2CuO_4 should be a good metal. The problem then becomes one of identifying the electron-electron interaction, ignored in the band structure calculations, that cause this electron localization. In view of the fact that stoichiometric La_2CuO_4 has a half filled $d_{x^2-y^2}$ band and is antiferromagnetic, it is tempting to consider it a Mott insulator at low temperatures despite the broad band picture from the band structure calculations. However, the reduced moment on Cu found from neutron diffraction studies suggests that a spin density wave may be a better description of the magnetic interaction which leads to electron localization in La_2CuO_4. Clearly a better understanding of the properties of semiconducting La_2CuO_4 is crucial to our understanding of superconductivity in the copper oxide systems.

Superconductivity has been reported for some samples of La_2CuO_4. Current evidence indicates that such samples deviate from the ideal stoichiometry. Furthermore, only a small portion of a given sample seems to exhibit superconductivity (19). Considerably more work is needed to understand the electrical and magnetic properties of $La_{2-y}CuO_{4-x}$ as a function of x and y.

Some evidence has been presented that $La_{2-x}A_xCuO_4$ phases may distort to a monoclinic structure at low

temperatures (20). Clearly if this distortion occurs,
it is very small (21). The required line splitting in
powder diffraction patterns has not been observed even
when using high resolution techniques. Furthermore, the
neutron diffraction refinements give entirely
satisfactory results when orthorhombic symmetry is
assumed. Thus the displacements of atoms from the
orthorhombic model would be very small and it is thus
unlikely that two chemically different copper positions
(e.g. Cu^I and Cu^{III}) are produced. Furthermore, this
evidence for distortion is found in metallic and
semiconducting materials. Thus, it apparently is not a
distortion to be associated with the localized electron
properties found in stoichiometric La_2CuO_4.

In the $La_{2-x}A_xCuO_4$ phases where A is Ca, Sr, or Ba,
the tetragonal-to-orthorhombic transition is suppressed
as x increases. This transition in fact disappears at
about x = 0.2 which is just about the same point that
superconductivity disappears. With increasing x, T_c
first increases reaching a maximum of about 40 K where x
is about 0.15. Then T_c decreases and superconductivity
disappears with increasing x. This suggests that there
may be something favorable for superconductivity in the
buckled CuO_2 sheets as opposed to the flat CuO_2 sheets
found in the tetragonal La_2CuO_4 structure. It should be
noted that $La_{2-x}A_xCuO_4$ phases can also be prepared
(22,23) where A is Cd^{II} or Pb^{II}. These phases are,
however, not superconducting and will be discussed
further in section 1.3.2.

The structure of R_2CuO_4 phases with R = Pr, Nd, Sm
and Gd (Figure 1.5) is very similar to that of La_2CuO_4,
especially tetragonal La_2CuO_4. Exactly the same
infinite sheets of CuO_2 exist where the copper is in
square planar coordination to oxygen. In both the

La_2CuO_4 and Pr_2CuO_4 structures, these CuO_2 sheets are
bound together by very ionic bonds to the rare earth
cation and additional oxygen. The Cu-O bonding in the
CuO_2 sheets is weaker in the Pr_2CuO_4 structure than in
La_2CuO_4. This is reflected by the larger Cu-O distance
(1.98 Å vs 1.89 Å). Through inductive effects (1), one
expects Cu-O bonds to become more ionic as the A cation
becomes less electropositive in an A_2CuO_4 series.
Apparently, we may conclude that the Cu-O bonding in the
R_2CuO_4 phases for R = Pr, Nd, Sm and Gd is too ionic to
develop bands that support metallic conductivity. This
conclusion is supported by results showing that even
samples of $Pr_{1.9}Sr_{0.1}CuO_4$ and $Nd_{1.9}Sr_{0.1}CuO_4$ possess
high electrical resistivity (22,23). If these A_2CuO_4
materials lacked metallic properties due to a filled
band rather than because of high ionic character, this
doping should have produced conducting materials. At
this laboratory we have also recently refined the
structures of Pr_2CuO_4 and Nd_2CuO_4 at 10 K and 298 K from
neutron diffraction data to confirm that the structures
are correct and that there is only one type of Cu in
this structure (24).

1.2.10 $RBa_2Cu_3O_x$ Phases

The structural feature common to both the 1:2:3
compounds (Figure 1.6) and La_2CuO_4 is the infinite CuO_2
sheets. In both structures, these sheets contain copper
in square planar coordination to oxygen. In both cases,
these sheets are buckled, but in different ways. In the
1:2:3 structure, the oxygen and copper atoms in the CuO_2
sheets are found in just two planes slightly displaced
from each other. In the orthorhombic La_2CuO_4 structure,
the oxygens of the CuO_2 sheet are alternately above and

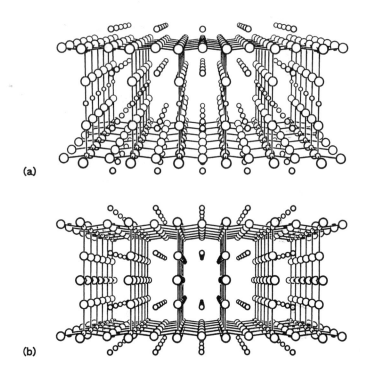

(a)

(b)

FIGURE 1.6   The structure of $YBa_2Cu_3O_6$ (a) and $YBa_2Cu_3O_7$
(b).   The small connected spheres are Cu.   The large
connected spheres are O.   The isolated spheres are Ba
(inside sandwich) or Y (outside sandwich).

below the plane of copper atoms.   However in both
structures, the final result is that the Cu-O-Cu bonds
are slightly bent from 180°.   In the La_2CuO_4 structure
the O-Cu-O bonds remain 180°, whereas this angle
deviates from 180° in the 1:2:3 structure.

   In the $YBa_2Cu_3O_6$ structure (25), two sheets of
$Cu^{II}O_2$ are bound together by an array of parallel
O-Cu^I-O linear sticks.   This brings the coordination of
copper in the $Cu^{II}O_2$ sheets up to five and forms a
sandwich containing two CuO_2 sheets and the array of
parallel O-Cu^I-O sticks.   The barium cations fit into

cavities within this sandwich. The rare earth cations are found between sandwiches, and they bind the sandwiches together to form a three dimensional structure from the two dimensional sandwiches.

This structural description has so far ignored the seventh oxygen. When the 1:2:3 structure forms, its composition is closer to $RBa_2Cu_3O_6$ than to $RBa_2Cu_3O_7$, and it is the structure of $RBa_2Cu_3O_6$ which has just been described. Semiconducting properties are found for $RBa_2Cu_3O_6$; thus, localized electrons and whole number oxidation states may be assumed. The formula for $RBa_2Cu_3O_6$ should be written as $RBa_2Cu^ICu_2^{II}O_6$ where Cu^{II} resides in the CuO_2 sheets and Cu^I resides in the O-Cu-O sticks. This two-fold linear coordination is preferred by Cu^I and is found in Cu_2O and many other oxides of copper.

When the $RBa_2Cu_3O_x$ phase is formed at 900° to 1000°C in air, the actual composition is about $RBa_2Cu_3O_{6.3}$. The extra 0.3 oxygen is highly mobile and tends to occupy at least two different lattice sites. Thus, no particular lattice site is more than 15% occupied by this oxygen. On cooling in air, additional oxygen is acquired and it tends to order on specific lattice sites. The site it picks is between two coppers of the O-Cu-O sticks. It could enter the lattice along both the a and b axes or it could order along one of these axes. In fact, it tends to order along one of the axes, and by convention this is now referred to as the b axis. This causes a drop in symmetry from tetragonal to orthorhombic. Presumably, the main driving force for the ordering of oxygens along one axis is that this produces the square planar environment very much preferred for Cu^{II} and Cu^{III}. A disordered arrangement would produce highly unfavorable coordinations $\overset{\text{O}}{\underset{\text{O}}{\mid}}$ Cu-O and $\overset{\text{O}}{\underset{\text{O}}{\mid}}$ Cu $\underset{\text{O}}{\overset{\text{O}}{<}}$ instead of favored $\overset{\text{O}}{\underset{\text{O}}{\mid}}$ O-Cu-O.

It is interesting that the placement of oxygen between two coppers causes an expansion along that axis (the *b* axis). This is the opposite of what one observes in ionic structures. In the ionic model, two cations are repelled by their like charges. Placing an anion between the cations pulls them together. The fact that oxygen pushes the coppers apart in the 1:2:3 structure is one of many indicators that the Cu-O bonding is highly covalent.

Going from $RBa_2Cu_3O_6$ to $RBa_2Cu_3O_7$ is then an intercalation reaction. Most compounds made by intercalation reactions cannot be made directly, and they are in fact not thermodynamically stable at any pressure or temperature. This is also apparently true for the $RBa_2Cu_3O_7$ superconductors. Despite the fact that these superconductors are thermodynamically unstable materials, they may well be kinetically stable at room temperature and below, as are window glass and diamond.

When the $RBa_2Cu_3O_6$ absorbs oxygen to form an ordered $RBa_2Cu_3O_7$ structure, linear $-(Cu-O)_n$ chains develop along the *b* axis. There has been considerable speculation about how these chains might cause such high-temperature superconductivity ($T_c \sim 95$ K). In fact, these chains are not infinite. They zig-zag (turn 90°) at intervals of about fifty to hundreds of angstroms. This 90° turn occurs at 110 twin boundaries. This twinning is present in essentially all samples of orthorhombic 1:2:3 compounds, and its origin is very easy to understand. When the 1:2:3 structure forms, it is tetragonal. As it cools and acquires more oxygen, the symmetry changes to orthorhombic. The a and b cell edges are no longer equal, and the crystal must undergo a shape change. It is well known from examples of ferroelectric and ferroelastic materials, that crystals resist changing their shape when undergoing distortions to lower symmetry. Multiple twinning allows the lowering of symmetry without a change in the external shape of the crystal.

It is interesting to examine the composition and atomic structure at the twin boundary. Oxygen deficiency at this boundary is likely and may explain why it is difficult to achieve an oxygen stoichiometry of exactly $RBa_2Cu_3O_7$. Typical preparations of the 1:2:3 phases are oxygen deficient, e.g. $RBa_2Cu_3O_{6.8}$. The oxygen vacancies may well have collected at twin boundaries, and the degree of oxygen deficiency would then be related to the concentration of twin boundaries. If the oxygen content is forced up to $RBa_2Cu_3O_7$, the copper atoms where two chains intersect will have an

unusual four-fold coordination, $Cu\underset{O}{\overset{O}{\langle}}$ instead of $O-Cu-O$.

This unfavorable situation may lead to peroxide formation (see below).

For $RBa_2Cu_3O_x$ phases, several changes tend to occur as x decreases from 7 to 6. The orthorhombic symmetry gives way to tetragonal symmetry, anywhere from x = 6.6 to x = 6.3 depending on the treatment conditions. The superconducting properties also degrade and disappear along with this change from orthorhombic-to-tetragonal symmetry (Figure 1.7). In fact, the materials become semiconducting. We now know that the structure and properties of $RBa_2Cu_3O_x$ phases are dependent on sample thermal history as well as the value of x. It appears

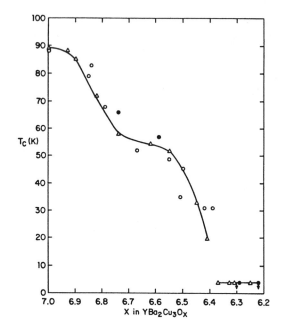

FIGURE 1.7   Effect of oxygen stoichiometry on T_c.
Samples obtained by quenching from high temperature.
(From Ref. 26.)

that $-(Cu-O)_n$ chains or chain segments along the b axis
are necessary for high-temperature superconductivity.
Such chains will normally, but not necessarily, result
in orthorhombic rather than tetragonal symmetry. For
$La_{1.25}Ba_{1.75}Cu_3O_{7.1}$ with the $YBa_2Cu_3O_7$ structure (some
La on Ba sites), we find bulk superconductivity with a
T_c of about 60 K even though the symmetry is clearly
tetragonal (27). Although one might conclude an absence
of chains for tetragonal symmetry, the chains are most
likely still present but randomly oriented along the a
and b axes. Thus, the average symmetry is tetragonal,
but the local symmetry would be orthorhombic.

There has been a general lack of success in lowering the temperature required to prepare 1:2:3 compounds. In this laboratory, we have prepared $YBa_2Cu_3O_x$ phases at 800°C, but they are not superconducting (28). Neutron diffraction and electron microscopy show a higher level of disorder than that seen in 1:2:3 materials prepared above 900°C. Although they are orthorhombic, the b/a ratio is close to one, and the $-(Cu-O)_n$ chains are apparently poorly formed.

Many structural refinements of 1:2:3 phases show an unusually large thermal parameter or disorder for the oxygen in the $-(Cu-O)_n$ chains. The usual result is that this oxygen is very unstable in the a-b plane but behaves in a normal manner with regard to the c axis. We find that this effect is exaggerated with increasing size of the R cation and in our low temperature preparation. There are various ways to model this disorder, and more work is required. However, results from our refinements at different temperatures indicate that this behavior is positional disorder rather than a true vibration. One positional disorder model which can account for the diffraction results is to assume that peroxide anions are tending to form when oxygen is in the "wrong chain" position. That is, the four-fold coordination to stabilize Cu^{II} or Cu^{III} should be square planar.

If Cu forms instead, this might convert to Cu.

This peroxide formation would reduce the copper and destroy superconductivity.

There have been many substitutions into the 1:2:3 structure. None of these have resulted in higher T_c's. It is now well known that the R cation in the $RBa_2Cu_3O_x$

phase can be yttrium or any lanthanide except Ce and Tb.
Presumably Ce and Tb have too great a preference for the
tetravalent state to form the 1:2:3 structure. Although
$PrBa_2Cu_3O_7$ can be formed with the 1:2:3 structure and
the Pr appears to be trivalent, superconducting
properties have not been observed. It would appear that
the mixed valency situation of Cu required for
superconductivity is disrupted by a $Pr^{III}-Pr^{IV}$ mixed
valency even though the Pr^{IV} content is very low.
Substitutions of Sr for Ba lead to a drop in T_c and an
eventual disappearance of superconductivity. The only
significant rare earth substitution for Ba is in the
case of La, i.e. $La_{1+x}Ba_{2-x}Cu_3O_y$. The T_c decreases with
increasing x. The upper limit of x is about 0.5 where
one has a tetragonal material which is semiconducting
and not superconducting (29). Substitutions for copper
have been complicated by the fact that there are two
different copper sites. We find from our neutron
diffraction results that Fe substitutes on both copper
sites. Substitution on the CuO_2 layer is not unexpected
because Fe^{III} is known to occur in such a five-fold
coordination site. When Fe substitutes into the copper
chain site, it presumably takes along sufficient oxygen
to form an octahedron. The substitution of Aℓ for
copper is interesting because it strongly prefers the
chain site (30). Presumably Aℓ also carries in extra
oxygen to make the chain site an octahedral site.
Replacing up to 10% of the chain coppers with Aℓ has
little effect on the superconducting properties. Higher
substitution levels cause a significant drop in T_c.

   There has been much discussion of anion
substitution for 1:2:3 phases. Fluorine is the only
anion one might expect to readily substitute directly
for the oxygen in the 1:2:3 structure. Indeed this

substitution probably does occur to some extent but with no improvement of superconducting properties. There is no known example of metallic conductivity in a metal fluoride where the conduction band is formed from metal-fluorine orbital overlap. Fluorides are simply too ionic. Thus, it is very disruptive for electron delocalization to replace oxygen with fluorine in the 1:2:3 lattice. Furthermore, higher T_c's in the copper oxide superconductors have been associated with higher average copper oxidation state. Replacing oxygen with fluorine would lower the average copper oxidation state.

1.3   MECHANISM

A consensus has developed that the classical BCS theory does not apply to the high temperature oxide superconductors: $Ba(Pb,Bi)O_3$, $(La,A)_2CuO_4$ and $RBa_2Cu_3O_7$. Nonetheless, the current evidence compels one to believe that the electrons responsible for superconductivity in these materials are paired with opposite spins. The classical BCS model depends on an electron-phonon coupling to produce the Cooper pairs. The low density of states at the Fermi level and the very small isotope effect on T_c for $YBa_2Cu_3O_7$ causes us to look for an electron-electron mechanism for producing the Cooper pairs necessary for the superconducting state. Before discussing a possible pairing mechanism, we must digress for some background on oxidation states and covalency.

1.3.1   Oxidation states vs. real charges

Inorganic chemistry could not be systematized without the concept of oxidation states. It is challenging to define precisely what one means by an oxidation state,

but this electron accounting scheme works exceedingly
well. It allows us to predict which combination of
elements is likely to form compounds. For compounds
that do form, the system of formal oxidation states
allows for a prediction of the number of unpaired
electrons and their character.

When we write $Cu^{II}O$, we do not mean that the real
charges on Cu and O are +2 and -2, respectively. The
bonding orbitals on Cu and O have strongly mixed to
produce new energy levels which contain strong
contributions from both copper and oxygen. Nonetheless,
writing $Cu^{II}O$ does imply that there has been some
electron flow from the copper atom to the oxygen atom.
Thus, it is convenient to refer to these two species as
cations and anions. Even though the real charge on
copper and oxygen is closer to one than two, we should
not be tempted to write CuO and $Cu^{I}O^{-I}$. The O^{-I}
radical species, if sufficiently concentrated, will
readily dimerize to form peroxide: $2O^{-I} \rightarrow 2O_2^{-II}$.
Furthermore, this peroxide species tends to be unstable
with respect to disproportionation and oxygen evolution:
$2O_2^{-II} \rightarrow 2O^{-II} + O_2 \uparrow$.

Even in highly ionic oxides, the real charges on
cations and anions are only about half the formal
oxidation state. Real charges in highly covalent
compounds are in fact closer to zero than to their
formal oxidation states. This is the case for Bi in
$Ba_2Bi^{III}Bi^{IV}O_6$ with the net effect that the real charge
difference between Bi^{III} and Bi^V is less than one.
Nonetheless, $Ba_2Bi^{III}Bi^VO_6$ is the correct way to
describe this compound where there are clearly two
different types of Bi cations ordered on the octahedral
sites of the perovskite structure. For a complete
discussion of the relationship between oxidation states

and real charges, the reader is referred to Sanderson's book "Chemical Bonds and Bond Energy" (31).

The $2p$ levels of oxygen for CuO and $BaBiO_3$ contain a large contribution from Cu and Bi, respectively. Likewise, the valence band closest to the Fermi level contains a large oxygen contribution even though we may label these bands as Cu $3d$ and Bi $6s$ bands. Such labels do correctly indicate the basic character of such bands.

It is well known that the O^{2-} species is not stable in the gas phase with respect to the reaction $O^{-2} \rightarrow O^{-1} + e^-$. However, in solids the divalent O^{-II} anion becomes stabilized because of covalency and because of the positive species, i.e. cations, that surround it. It is important to recognize that the more electropositive cations, such as Ba^{2+}, are more effective in stabilizing the $2p$ orbitals of oxygen. Thus Cu^{III} and Bi^V are oxidation states that are not stable without the presence of very basic cations that stabilize the oxygen $2p$ states sufficiently to allow Cu^{III} and Bi^V to coexist with O^{-II}.

When examining the number of cation (Cu or Bi) states $vs.$ oxygen states near the Fermi level, one must keep in mind that these oxides are oxygen rich. Thus, the high number of oxygen states in the "metal d or s band" is directly related to covalency and directly related to the relatively large number of oxygens present.

1.3.2 Disproportionation Mechanism

$Ba(Pb,Bi)O_3$

For $BaBiO_3$ itself, semiconducting properties are observed because disproportionation has occurred and this compound should be written as $Ba_2Bi^{III}Bi^VO_6$. The two different Bi cations are ordered on the octahedral sites of the perovskite structure just as in the case of Ba_2MgWO_6. Normally perovskites of the $Ba_2M^{III}M^VO_6$ type do not show complete ordering of the octahedral site cations. However, in $BaBiO_3$, complete order is easily achieved because no cation diffusion is required to completely order Bi^{III} and Bi^V; only electrons need order. $BaBiO_3$ may be described as having a charge density wave, but this is equivalent to the description just given.

BaPbO$_3$ is best described as a semimetal (Figure 1.2). As Bi is substituted for Pb, i.e. $BaPb_{1-x}Bi_xO_3$, the electron concentration in the 6s band increases. To the extent that formal oxidation states have meaning in metallic phases, we may refer to bismuth as Bi^{IV}. However, the 6s electron of Bi^{IV} is delocalized in the 6s band. As the Bi concentration increases, the Bi^{IV} cations begin to interact. The final effect of the interaction is to disproportionate into Bi^{III} and Bi^V. This is then an electron pairing interaction which could result in the observed superconductivity. As the Bi concentration is further increased, the tendency to pair localizes the 6s electrons as pairs, i.e. Bi^{III} formation. Thus, we have semiconducting properties. The empty 6s band and the filled O2p band no longer overlap to give metallic conductivity, at least partly because the lattice has expanded with the Bi substitution.

For the $BaPb_{1-x}Bi_xO_3$ system, T_c reaches its maximum just when the disproportionated state (Bi^{III} + Bi^V) becomes nearly degenerate with the nondisproportionated state (Bi^{IV}).

Systems with d^9 Cations

The analogous disproportionation reaction for
superconductivity in the copper oxides would be:
$2Cu^{II} \rightarrow Cu^I + Cu^{III}$. Such disproportionation is not
known for Cu^{II}, but it is well known for the other d^9
cations, Ag^{II} and Au^{II}. The tendency for the d^9 cations
to disproportionate increases going down the
$Cu^{II} \rightarrow Ag^{II} \rightarrow Au^{II}$ column of the periodic table.
Presumably this increased tendency to disproportionate
is caused by increased covalency of the bonding as one
goes down this column of the periodic table (32). With
the d shell nearly filled, the antibonding interactions
are very significant and of course become more
significant as the covalency increases. In the ionic
limit, the bonding and antibonding interactions are
merely symmetry considerations. However, as covalency
increases, there are several ways to reduce the
antibonding interaction. The bending of M-O-M bonds to
reduce the π^* interactions has already been discussed
(section 1.2.9).

The σ^* antibonding interaction cannot be alleviated
as simply as the π^* interaction. However, the σ^*
interaction can be alleviated by decreasing the cation
coordination number. For d^{10} cations, there is a
tendency for two-fold linear coordination, common for
Cu^I, Ag^I and Hg^{II}, until the d shell has sufficiently
contracted as in Zn^{II} so the sp bonds can form without
significant antibonding interference from the d shell.
In effect, hybridization of the d shell with the s and p
orbitals polarizes the filled core away from the two
ligands, thus allowing the formation of just two bonds.
For the d^8 and d^9 cations, there is a driving force
towards square planar coordination. With the filled d_{z^2}

orbital oriented perpendicular to the plane of the
bonds, a significant sigma antibonding interaction is
avoided (Cu^{III}) or diminished (Cu^{II}). For both the
two-fold coordination of d^{10} cations and the square
planar coordination for d^8 and d^9 cations, the essential
feature is that we polarize the cation core away from
the bonds to reduce antibonding interactions.

For the proposed disproportionation mechanism for
superconductivity, it is essential that the d^9 cation
not be in a symmetry which would lead to a degeneracy of
the half occupied state with any other state. Thus
octahedral coordination for Cu^{II} is not conducive to
superconductivity. The d_{z2} and d_{x2-y2} orbitals are then
degenerate, and the disproportionation reaction is no
longer simply spin pairing. This degeneracy is removed
for Cu^{II} in La_2CuO_4 and $YBa_2Cu_3O_7$.

A major challenge for chemists is to adjust the
covalency of the Cu^{II}-O bonds. In the ionic limit, the
$3d$ electrons of Cu^{II} will be localized. In the covalent
limit, disproportionation to Cu^I and Cu^{III} will occur.
Thus the two limits give insulating materials (Figure
1.8). An intermediate metallic state may occur in cases
where there are infinite (Cu-O) connections.
Superconductivity may also occur if the degeneracy
between d_{z2} and d_{x2-y2} has been removed.

The way to vary the covalency of Cu^{II}-O bonds is
through inductive effects (1). The addition of strongly
electropositive cations such as Ba^{II} is a route to
increasing the covalency of Cu^{II}-O bonds. In the R_2CuO_4
series, the important structural feature of square
planar CuO_2 sheets is maintained for R = La, Pr, Nd, Sm
and Gd. The electronegativity of R decreases steadily
on going from Gd to La. Thus the Cu-O bonds become more
covalent. For R = Pr, Nd, Sm and Gd, the Cu^{II}-O bonds

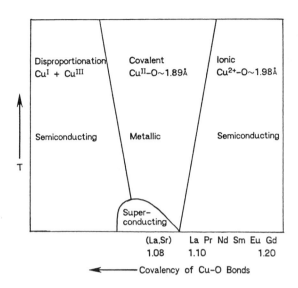

FIGURE 1.8 Schematic phase diagram for R_2CuO_4 phases where R = Gd, Sm, Nd, Pr, La or $La_{0.9}Sr_{0.1}$. The electronegativity values are given for the R cations.

are sufficiently ionic that semiconducting properties are observed. We know that this is an ionic situation instead of a filled band situation because doping these phases with divalent cations, i.e. $R_{2-x}Sr_xCuO_4$ does not produce conducting materials. The covalency of $Cu^{II}-O$ bonds in La_2CuO_4 has increased to the level where metal-like properties are observed at room temperature and above. There has been a structure change relative to the other R_2CuO_4 phases with regard to the placement of oxygens between the CuO_2 sheets, but the sheets themselves are unchanged in tetragonal La_2CuO_4 except that the Cu-O distances are shorter, consistent with the higher covalency. The distortion to orthorhombic La_2CuO_4 is a direct result from this higher covalency, as previously discussed (section 1.2.9).

Small portions of some "La_2CuO_4" samples can be superconducting. However, true bulk superconductivity apparently requires the substitution for La of Ba, Sr or Ca. It should be noted that these cations are all more electropositive than La. Thus they all cause increased covalency of Cu-O bonds coupled with decreased Cu-O distances. In fact, it has been observed that T_c increases in a regular way with decreasing Cu-O distance (33). This distance may be decreased by increasing x in $La_{2-x}A_x^{II}CuO_4$ where A is Ba, Sr or Ca. Also, increasing pressure results in a decreased Cu-O distance and increased T_c.

Superconductivity is not produced in $La_{2-x}A_x^{II}CuO_4$ phases when A is Pb or Cd. The reasons for this are presumably related to the fact that Pb^{II} and Cd^{II} are less electropositive than La^{III}. This could then cause an increased ionicity of the Cu-O bonds. It could also make it more difficult to stabilize Cu^{III}.

Series of the phases of the type $(La,R,Sr)_2CuO_4$ have been prepared in several laboratories (34). For samples where the Sr content is fixed, it is observed that R substitution for La causes a decreased T_c. It was suggested (34) that a unit cell contraction caused this decrease in T_c, but this explanation is inconsistent with other observations that T_c increases as the Cu-O distance decreases. It seems more likely that the drop in T_c is related to the more electronegative character of the R cations relative to La. This would be expected (1) to increase the ionicity of Cu-O bonds and therefore decrease T_c. Furthermore, T_c drops in a regular manner as the electronegativity of R increases.

Systems with d^1 Cations

We need to examine the possibility of whether or not the
disproportionation mechanism for superconductivity
should be applied to systems with d^1 cations where the
conduction band is π^* in character instead of σ^* as it
is for the bismuth and copper oxide superconductors.
The problem with applying it to the d^1 systems is a
degeneracy of the π levels if the d^1 cation is in high
symmetry such as octahedral. For octahedral
coordination to oxygen, the d_{xy}, d_{zx}, and d_{zy} levels are
degenerate and contain just one electron for the d^1
case. The disproportionation reaction in this situation
does create a paired state but the electrons in the
paired state would have the same spin. In order to have
opposite spins in the paired state, we must remove the
degeneracy from the d_{xy}, d_{zx}, and d_{zy} orbitals. One way
to remove this degeneracy is through ferroelectric type
distortions which are common for perovskite type
structures where the d shell is empty or nearly so. We
should also remember that superconductivity never occurs
in cubic A_xWO_3 or A_xMoO_3 bronzes. In the hexagonal and
tetragonal tungsten bronzes, the d_{xy}, d_{zx}, and d_{zy}
orbitals would not be degenerate. Also in the spinel
structure found for $LiTi_2O_4$, the site symmetry for Ti is
not strictly octahedral; thus, the d_{xy}, d_{zx}, and d_{zy}
orbitals are not degenerate. We must therefore conclude
that there is good reason to suspect that the
disproportionation based mechanism for superconductivity
may apply to the d^1 situation as well as to s^1 and d^9
situations.

1.4 DISCUSSION

It appears necessary for superconductivity to dilute the disproportionating species with the associated empty state, i.e. s^1 in s^0, d^1 in d^0, and Cu^{II} in Cu^{III}. The reasons for this are not entirely clear but probably generally relate to phenomena that compete with superconductivity such as charge density waves. Clearly the disproportionation mechanism for superconductivity does not apply unless we defeat the tendency to localize electrons in charge density waves or spin density waves. One way to defeat this is by dilution in just the right matrix. Another way is through frustration which seems to be a factor in at least the La_2CuO_4 based superconductors [1]. Still another way to defeat such localization is through defects. The substitution of A^{II} for La^{III} in La_2CuO_4 may serve this purpose. The oxygen defects in the $RBa_2Cu_3O_{7-x}$ may also be important in this capacity.

The disproportionation mechanism for superconductivity as presented here is very limited in its application. It only applies to d^1, d^9 and s^1 situations. For the d^1 and d^9 situation, we must depart sufficiently from cubic symmetry to remove degeneracies of the d levels. There is of course no reason why this mechanism should be restricted to oxides. One would expect that some sulfides, for example, should be candidates for this type of superconductivity.

An important issue for the disproportionation mechanism for superconductivity is how this mechanism relates to lattice vibrations and thus the classical BCS theory based on electron-phonon interactions. Although in principle the mechanism proposed here does not require an electron-phonon interaction to produce superconductivity (35), one would always expect some coupling between conduction electrons and oxygen vibrations. Thus we would expect some shifts of T$_c$ when ^{16}O is replaced with ^{18}O. In fact, such a shift is found for all oxide superconductors (36). Calculation of the magnitude of this shift is not straight forward, and this is currently an area of considerable activity for theorists.

ACKNOWLEDGMENTS

I am grateful to W. Bindloss and A. Suna for discussions, to U. Chowdhry for a critical reading of this paper, and to C. C. Torardi for the structure drawings.

REFERENCES

These are very exciting times, but we are presented with some unique problems. One of them is how to refer to the work of others. Much of what we know is from preprints and by word-of-mouth. Also, many different laboratories have obtained essentially the same results; it is very difficult to know who was first. Thus one finds that one is dealing with a large number of poorly defined references and large author lists. The reference section could easily become the largest section of the paper. To avoid this situation, I make reference only in those situations where some critical point is being made.

1. A. W. Sleight, High-Temperature Superconductivity in Oxides, in Chemistry of High-Temperature Superconductors (D. L. Nelson, M. S. Whittingham,

T.F. George, eds.), American Chemical Society
Symposium Series, Washington D.C., 1987, pp. 2-12.

2. A. W. Sleight, Chemtronics, 2, 116 (1987).

3. J. K. Hulm, C. K. Jones, R. Mazelsky, R. A. Hein,
and J. W. Gibson, Proc. 9th Int. Conf. on Low Temp.
Physics (J. G. Daunt, D. O. Edwards, F. J. Milford,
M. Yaqub, eds.), 600 (1965).

4. J. J. Schooley, W. R. Hosler, and M. L. Cohen, Phys.
Rev. Lett., 12 474 (1964).

5. A. R. Sweedler, C. Raub, and B. T. Matthias, Phys.
Lett., 15 108 (1965).

6. A. W. Sleight, T. A. Bither, P. E. Bierstedt, Solid
State Commun., 7 299 (1969).

7. A. W. Sleight, and J. L. Gillson, Solid State
Commun., 4 601 (1966).

8. D. C. Johnston, H. Prakash, W. H. Zachariasen, R.
Viswanathan, Mater. Res. Bull., 8 777 (1973).

9. A. W. Sleight, J. L. Gillson, P.E. Bierstedt, Solid
State Commun., 17 27 (1975).

10. B. Batlogg, in Physics Today, 40 23 (1987).

11. V. V. Bagotko and Yu. N. Venevtsev, Sov. Phys. Solid
State, 22 705 (1980).

12. D. E. Cox and A. W. Sleight, Solid State Commun., 19
969 (1976); D. E. Cox and A. W. Sleight, Acta
Cryst., B35 1 (1979).

13. G. Thornton and A. J. Jacobson, Acta Cryst., B34 351
(1978).

14. A. W. Sleight and D. E. Cox, Solid State Commun., 58
347 (1986).

15. D. E. Cox and A. W. Sleight, Proc. Conf. on Neutron
Scattering, Gathinburg, Tennessee, (R. M. Moon, ed.)
National Technical Information Service, Springfield,
VA, pp. 45-54.

16. B. Batlogg, J. P. Remeika, R. C. Dynes, H. Barz, A. S. Cooper and J. P. Garno, <u>Proc. Conf. on Superconductivity in d- and f-Band Metals</u>, Karlsruhe (1982) p. 402.

17. M. B. Robin, K. Andres, T. H. Geballe, N. A. Kuebler and D. B. McWhan, <u>Phys. Rev. Lett.</u>, <u>17</u> 917 (1966).

18. R.J.D. Tilley and A. H. Davies, <u>Nature</u>, <u>326</u> 859 (1987).

19. P. M. Grant, et al, <u>Phys. Rev. Lett.</u>, <u>58</u> 2482 (1987).

20. S. C. Moss, K. Forster, J. D. Axe, H. You, D. Hohlwein, D. E. Cox, P. H. Hor, R. L. Meng and C. W. Chu, <u>Phys. Rev.</u>, <u>35B</u> 7195 (1987).

21. A very large monoclinic distortion for La_2CuO_4 at very low temperatures was suggested [E. F. Skelton, et al. <u>Phys. Rev.</u>, <u>B36</u> 5713 (1987)], but the observed x-ray diffraction pattern was subsequently identified as solid nitrogen.

22. I. S. Shaplygin, B. G. Kakhan, and V. B. Lazarev, <u>Russ. J. Inorq. Chem.</u>, <u>24</u> 820 (1979).

23. J. Gopalakrishnan, M. A. Subramanian, and A. W. Sleight, to be published.

24. C. C. Torardi, M. A. Subramanian, and A. W. Sleight, to be published.

25. C. C. Torardi, E. M. McCarron, P. E. Bierstedt, and A. W. Sleight, <u>Solid State Commun.</u>, in press.

26. W. E. Farneth, R. K. Bordia, E. M. McCarron, M. K. Crawford, and R. B. Flippen, to be published.

27. E. M. McCarron, C. C. Torardi, J. P. Attfield, K. J. Morrissey, A. W. Sleight, D. E. Cox, R. K. Bordia, W. E. Farneth, R. B. Flippen, M. A. Subramanian, F. Lodrup, and S. J. Poon, Materials Research Society Proceedings, Boston, MA (1987).

28. C. C. Torardi, E. M. McCarron, M. A. Subramanian, H.
    S. Horowitz, J. B. Michel, A. W. Sleight, and D. E.
    Cox, Structure-Property Relationships for $RBa_2Cu_3O_x$
    Phases, in Chemistry of High-Temperature
    Superconductors (D. L. Nelson, M. S. Whittingham,
    T.F. George, eds.), American Chemical Society
    Symposium Series, Washington D.C., 1987, pp.
    152-163.

29. C. C. Torardi, E. M. McCarron, M. A. Subramanian,
    A. W. Sleight and D. E. Cox, Mat. Res. Bull., 22
    1563 (1987).

30. T. Siegrist, L. F. Schneemeyer, J. V. Waszczak, N.
    P. Singh, R. L. Opila, B. Batlogg, L. W. Rupp, and
    D. W. Murphy, Phys. Rev. B., in press.

31. R. T. Sanderson, Chemical Bonds and Bond Energy,
    Academic Press, N.Y. (1976).

32 Alternately, one may view the increased tendency for
    disproportionation in the Cu^{II}, Ag^{II}, Au^{II} series as
    due to increased involvement of the respective d
    shell.

33. K. Kishio, K. Kitazawa, N. Sugii, S. Kanbe, K.
    Fueki, H. Takagi, and S. Tanaka, Chem. Lett., 635
    (1987).

34. J. M. Tarascon, L. H. Greene, W. R. McKinnon, and G.
    W. Hull, Solid State Commun., 63, 499 (1987).

35. J. E. Hirsch and D. J. Scalapino, Phys. Rev., 32B
    5639 (1985).

36. H.-C. L. Loye, K. J. Leary, S. W. Keller, W. K. Ham,
    T. A. Faltens, J. N. Michaels, and A. M. Stacy,
    Science, in press.

2

The Advent of High-Temperature Superconducting Materials: Chronology of Events and Hallmark Developments

WILLIAM E. HATFIELD University of North Carolina at Chapel Hill,
Chapel Hill, North Carolina 27599

1 DISCOVERY OF SUPERCONDUCTIVITY

It has been known from the earliest experiments on the electrical
properties of conductors that the electrical resistivity of a
metal decreases when it is cooled. At 20 to 30 K the resistance
usually becomes constant at a value determined by the purity of
the metal and the crystalline perfection of the specimen being
measured. When metals are cooled to even lower temperatures some
of them exhibit an abrupt drop in resistivity and enter a state
in which there is no resistance to the flow of current. This new
state was first discovered in mercury by H. Kammerlingh Onnes at
the University of Leiden in the spring of of 1911 (1). Kammerlingh Onnes noted that upon cooling below 4.2 K, "Mercury has
passed into a new state, which on account of its extraordinary
properties, may be called a superconductive state".

The temperature at which the transition to the superconducting state occurs is designated T_c. More than 25 metallic

elements superconduct at ambient pressure, and several more do
when placed under high pressure. Transition temperatures for
some representative superconducting elements are given in Table
1 (2). There are several thousand superconducting alloys (3),
and the highest transition temperature achieved before 1986 was
23.2 K for Nb_3Ge (4). In late 1986 and early 1987 there were
tremendous advances made in the discovery and characterization of
new high temperature superconducting materials, and there are now
reports of superconducting transitions above room temperature.

The chronology of events and hallmark developments that led
to the advances in high temperature superconductivity are briefly
sketched in this tutorial paper. By necessity, certain terms are
briefly defined and fundamental concepts are tightly skectched as
they are encountered in the chronology. More detailed treatments
of the physics of superconductors and the structural and electronic properties of high temperature superconducting materials are
presented in the following papers by J. H. Miller and
M.-H. Whangbo.

2 THE SIGNATURE OF THE SUPERCONDUCTING STATE

Loss of Resistivity An abrupt drop in resistivity of a metal at
a characteristic temperature T_c and an apparent absence of
resistance to the flow of electricity below T_c, as shown in
Figure 1a, are dramatic indications that the superconducting
state may be present. This was the discovery made by Kammerlingh

Table 1. Superconducting Transition Temperatures for Some
Representative Elements

Element	T_c, K	Element	T_c, K
Aluminum	1.75	Cadmium	0.517
Indium	3.408	Iridium	0.113
Lead	7.196	Mercury-α	4.154
Molybdenum	0.915	Niobium	9.25
Osmium	0.66	Protactinium	1.4
Rhenium	1.697	Ruthenium	0.49
Tantalum	4.47	Technetium	7.8
Thallium	2.38	Thorium	1.38
Tin	3.722	Titanium	0.40
Tungsten	0.0154	Vanadium	5.40
Zinc	0.85	Zirconium	0.61

Onnes in 1911. There are problems with the experimental work.
Below T_c, R = 0, and from Ohm's law, V = IR, then V is also equal
to zero. Very low, but finite, resistivities are difficult to
determine in the laboratory, where, typically, the voltage is
measured in a circuit at a constant applied current, and the
resistance and resistivity are calculated. Since thermoelectric
effects affect the measurement of vanishingly small voltages, it
is easy to see why the measurement of very small resistivities is
subject to many sources of experimental error.

The complete disappearance of resistivity may be demonstrated by making use of the properties of rings. Consider that a
ring, as shown in Figure 2, is exposed to an external magnetic
field of flux density $\underline{B}$, which is changing as a function of

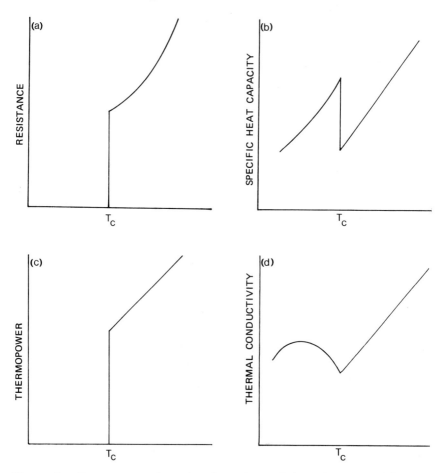

Figure 1. Temperature dependencies of the (a) resistance, (b)
specific heat, (c) thermopower, and (d) thermal conductivity of a
substance which undergoes a transition to a superconducting state
at T_C.

time. The current $\underline{I(t)}$ flowing in the ring at time $\underline{t}$ is given by Lenz's law, which states that

$$-A(dB/dt) = RI(t) + L[dI(t)/dt] \tag{\underline{1}}$$

where $\underline{A}$ is the area of the ring, $\underline{R}$ is the resistance, and $\underline{L}$ is the inducatance of the ring. If there is no external applied magnetic field, then ($\underline{1}$) becomes the differential equation

$$RI(t) + L[dI(t)/dt] = 0 \tag{\underline{2}}$$

which has the solution

$$I(t) = I(0)\exp(-Rt/L) \tag{\underline{3}}$$

Thus, the current in the ring exponentially decays in the absence of an external changing magnetic field, and it eventually vanishes.

However, if the ring in Figure 2 becomes superconducting, then $\underline{R} = 0$ and $\underline{I(t)} = \underline{I}(0)$. In other words, the current does not decay with time! This current is called the persistent current, and the circulation of the current gives rise to a magnetic field. Persistent currents remain without change, within the limits of detection by the most sensitive modern instruments, for many years. This property of superconducting materials is very important for high-field magnets such as those used in NMR spectrometers.

Abrupt Change in Specific Heat   The abrupt increase in specific heat at T_c that occurs for superconductors is shown in Figure 1b where it may be seen that the temperature dependence of the specific heat in the superconducting state is very much different from that in the normal state. The electronic specific heat

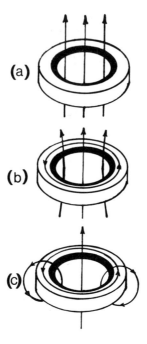

Figure 2. Schematic representation of a superconducting ring in
(a) the normal state in a magnetic field showing some flux lines,
(b) in the superconducting state with a current shown in the
ring, and (c) in the superconducting state with the persistent
current indicated.

above T_c is given by

$$C_{en} = \gamma T \qquad\qquad (\underline{4})$$

and that well below T_c is described well by

$$C_{es} = \gamma T_c a \exp(-bT_c/T) \qquad\qquad (\underline{5})$$

where $\underline{a}$ and $\underline{b}$ are experimental constants of about 10 and 1.5,
respectively. The exponential dependence of the specific heat in
the superconducting state suggests an average excitation energy

of $1.5kT_c$, or in other words, the specific heat results suggest
the existence of a gap in the electronic structure. The success
of the theory by Bardeen, Cooper, and Schrieffer (BCS theory) (5)
in describing this gap and the other fundamental properties of
superconductors will be discussed in the next paper in these
proceedings.

The existence of a gap has been confirmed spectroscopically
(6), and found to be about twice that of the thermal gap (6b,c).
This may be understood if the electrons occur as pairs with the
spectroscopic result giving the average energy required to create
a pair of excitations while the thermal result gives the energy
per statistically independent particle.

Thermopower and Thermal Conductivity If a metal in the normal
state is subjected to a temperature gradient δT, an electric
field $\underline{E}$ will be developed with the thermopower being defined as
$\delta T/E$. As shown in Figure 1c, the thermopower of a superconductor
disappears at T_c. This property may be exploited in devices, say
as switches, when superconducting materials with high T_c's become
available.

There is also an abrupt change in thermal conductivity at T_c
(Figure 1d). In some superconductors the thermal conductivity
increases as the temperature decreases below T_c, while in other
superconductors the thermal conductivity decreases with temperature below T_c. Determination of this property for specific
superconductors yields significant information concerning the
mechanism of the conduction process.

3 PERFECT DIAMAGNETISM OF SUPERCONDUCTORS, THE MEISSNER EFFECT

In 1933, Meissner and Ochsenfeld (7) reported that, upon cooling,
a magnetic field is expelled from a normal metal specimen when it
passes through T_c and becomes superconducting. This is the
Meissner effect. The exclusion of magnetic lines of flux from an
object in the superconducting state is shown schematically in
Figure 3, where a penetration depth, designated as λ, is indicated. Typical values for λ are on the order of 50 nm.

The intensity, $\underline{B}$, of a magnetic field along a specific
direction $\underline{i}$ within a body is given by

$$B_i = H_i + 4\pi M_i \tag{6}$$

where $\underline{H}$ is the applied magnetic field and $\underline{M}$ is the magnetization. Upon dividing through by H_i, there results

$$B_i/H_i = 1 + 4\pi(M_i/H_i) \tag{6a}$$

where the ratio M_i/H_i is the susceptibility of the body towards

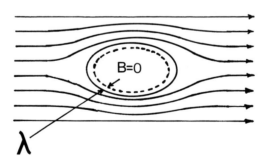

Figure 3. An object in the superconducting state in a magnetic
field. The penetration depth, λ, is indicated by the dashed
line.

induction in a field of strength H_i. The ratio is the volume
magnetic susceptibility and it is denoted by X. When placed in
an inhomogeneous magnetic field, a diamagnetic substance will
tend to move toward the weakest region (lowest density of
magnetic lines of force) of the inhomogeneous field, while a
paramagnetic substance will tend to move toward the strongest
region of the magnetic field.

Since $\underline{B}$ = 0 for superconductivity, then the lowest energy of
the system is achieved when the superconducting body is in the
weakest region of the magnetic field. The volume of the bulk
superconducting sample is much larger than the volume of the
penetration shell, $\underline{B}$ = 0, and

$$X_i = -1/4\pi \qquad\qquad\qquad (\underline{7})$$

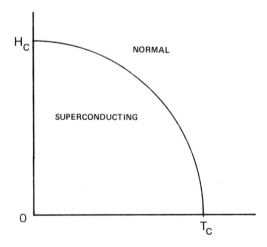

Figure 4.   The phase diagram for a superconductor which shows the
variation of the critical field H_c with temperature.

Thus, perfect superconductors exhibit <u>perfect diamagnetism</u> with a value of $-1/4\pi$.

The Meissner effect is reversible, and the temperature of the transition from the normal state to the superconducting state is dependent on the strength of the applied magnetic field. The critical field $\underline{H}_c$ is related to the thermodynamic energy difference between the normal and superconducting states. This energy difference is called the <u>condensation</u> energy of the superconducting state. The temperature dependence of the critical field is given by

$$H_c(T) = H_c(0)[1 - (T/T_c)^2] \tag{\underline{8}}$$

and the phase diagram traced out by Equation (8) is shown in Figure 4, where it may be seen that the critical field $H_c = 0$ at $T = T_c$. The transition at zero field is second order, but the transition in the presence of an applied magnetic field is first order, since there are discontinuities in state properties as well as an associated latent heat.

In superconducting substances, a few elements and most alloys and compounds exhibit an incomplete Meissner effect and have broad magnetic transitions. These properties may be understood in terms of the theoretical developments which were evolving simulataneously with the experimental advances.

4   THEORETICAL DEVELOPMENTS

<u>London Theory</u>   The discovery of the Meissner effect was soon followed by the work of F. London and H. London (8), who proposed

that the flux density $\underline{B}_x$ at a point $\underline{x}$ within a superconductor is given by

$$d^2 B(x)/dx^2 = B(x)/\lambda_L^2 \tag{9}$$

where λ_L is the London penetration depth. The solution of the differential equation in one dimension gives

$$B(x) = B(0) \exp(-x/\lambda_L) \tag{10}$$

which shows that the flux density decreases exponentially within the surface layer and essentially disappears when $\underline{x}$ is much greater than λ_L, a result which accounts for the Meissner effect. The London penetration depth is given by

$$\lambda_L = (mc^2/4\pi N_s e^2)^{1/2} \tag{11}$$

If N_s is taken to be 9×10^{21} e/cm^3, a commonly accepted value for the number density of conduction electrons, λ_L is on the order of 10^{-6} cm. Values on this order of magnitude are obtained for many superconductors in weak magnetic fields.

Non-local Generalization of London Theory   The London theory was unable to account for the properties of a number of superconductors, and it was necessary to modify the theory. Pippard (9) presented a successful model which is based on the uncertainty principle and the energy distribution of the electrons. The model recognizes that only those electrons with energy within kT_c of the Fermi energy will be involved in phenomena that occurs at T_c, and that those electrons have the Fermi velocity v_F. Thus, from the uncertainty principle, the range of momenta is $\delta p = kT_c/v_F$, and the range in position, or the characteristic length of the wavepacket ξ_o, is given by

$$\xi_o = ahv_F/2\bar{\pi}kT_c \tag{\underline{12}}$$

where a is a constant of the order of unity and must be determined experimentally. Experiments on tin and aluminum (10) yield a value for a of 0.15. BCS theory yields a value of 0.18 for the parameter a.

Two Types of Superconductors   It is possible to conclude immediately from the discussion in the previous sections that there are two types of superconductors, one type which may be described by the nonlocalized modification of the London model, and a second type by the original localized London model. These are called Type I and Type II superconductors, respectively, and they may be differentiated by their properties in an applied magnetic field. It is most revealing to consider the two types of superconductors in view of the additional theoretical developments presented in the following sections.

Ginzburg-Landau Theory and Abrikosov's Reversal   In 1950 Ginzburg and Landau (11) presented a hybrid quantum mechanical-phenomenological treatment based on a complex wave function for the superconducting electrons utilizing the variational principle. The theory introduced a temperature-dependent coherence length $\xi(T)$, which is the same as the Pippard coherence length far from T_c, but which diverges at T_c. The theory was successful since it provided an explanation for the intermediate state of Type I superconductors which will be discussed below.

For typical pure superconductors, λ_L is approximately 50 nm, is about 300 nm, and the ratio $K_{GL} = \lambda/\xi$, which is a small

number, is the Ginzburg-Landau parameter. The surface energies
implied by these dimensions are important in defining the domains
of superconducting and normal phases and the intermediate state.
The difference in surface energy between the superconducting and
normal state is on the order of $\xi H_c^2/8\bar{\pi}$, while the diamagnetic
energy loss is $\lambda H_c^2/8\bar{\pi}$. The positive surface energy results in a
domain pattern for the intermediate state of a Type I superconductor with dimensions ranging from the microscopic coherence
length ξ and the macroscopic sample size.

Arbikosov (12) considered the reverse situation in which $\lambda >$
ξ, and χ is large. It is clear that the opposite situation from
that described in the preceding paragraph must arise. That is,
the subdivision into domains proceeds until it is limited by the
microscopic length ξ. Superconductors which exhbit this property
are called Type II superconductors. There are two critical
fields for these superconductors. Consider the case of an
initially superconducting body in an increasing applied magnetic
field. At the critical field H_{c1} there is the onset of flux
penetration, and this flux penetration increases with the applied
magnetic field until a second critical field H_{c2} is reached. At
values of the applied field above H_{c2} the body becomes normal.
Since the energy required to exclude the magnetic field is
reduced as a result of partial flux penetration, then Type II
superconducing materials may have large values of H_{c2}. This
property has permitted the utilization of superconducting
materials in high-field solenoids and magnets.

Cooper Pairs   Interactions between electrons in a metal in the
normal state are usually ignored, and the resulting free electron
model may be used to describe a wide range of properties of
metals. Interactions between electons can not be ignored in
superconducting state. In the superconducting state some of the
electrons are bound together in pairs known as Cooper pairs
(13). The Cooper pairs are governed by certain requirements of
quantum mechanics, the most important of which may be the
requirement that all of the Cooper pairs have the same value of
total momentum. It may be shown that this property leads to the
static electromagnetic properties of zero resistance, Meissner
effect, and others. Other unpaired electrons can exist simultaneously with the Cooper pairs in the superconducting state, but
the unpaired electrons are like electrons in the normal state.
Currents carried by them are resistive, and their contributions
to the magnetic susceptibility are very small. A Cooper pair is
more stable than two unpaired electrons by the amount of the
binding energy. This is energy of the superconducting gap
introduced above.

Theory assumes that all of the electrons in a superconducting
body are paired into Cooper pairs at 0 K in the absence of an
applied magnetic field and an electric current. The Cooper pairs
are broken up when energy is supplied to the metal, for examples,
by application of an external magnetic field or by increasing the
temperature. The population of unpaired electrons is
proportional to $\exp(-E/kT)$. This provides an explanation for

many of the properties of superconducting substances, such as the
exponential dependence of the specific heat, described above.

It is now possible to present a simple picture of a mechanism for superconductivity and the Cooper pairs. Consider one
electron which is moving through the lattice of positive ions.
The negative field of the electron polarizes and distorts the
lattice of positive ions shown in Figure 5a to yield a region of
excess positive ions (Figure 5b) which, in turn, attracts the
second electron and yields an effective attractive interaction
between the two electrons. If the attractive interaction is
greater than the Coulomb repulsion, then superconductivity
results. The role of electron-lattice interactions in the
mechanism for superconductivity was first investigated by
Fröhlich (14), and the suggestion was confirmed experimentally by
the demonstration of an isotope effect on T_c and H_c (15).

An explanation may also be presented for the long lifetime
of the persistent current. Since the flux is quantized, the
current cannot decrease by an infinitesimal amount, but only by a
quantized amount. Furthermore, superconductivity is a collective
phenomenon, and the wave function involves all of the Cooper
pairs. Thus, the quantized jump requires a simulataneous event
of about 10^{20} pairs. Such an event is extremely improbable, and
there is no decay of the persistent current.

Bardeen, Cooper, Schrieffer (BCS) Theory   The theory of superconductivity proposed by Bardeen, Cooper, and Schrieffer (5) in
1957, and which has been extended and refined by numerous

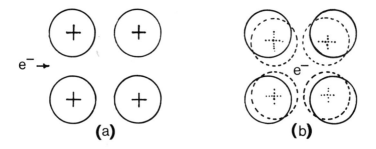

Figure 5. A simple model which shows the attraction of an
electron to a lattice distortion.

subsequent studies, has been successful in explaining essentially

all of the phenomena associated with the superconducting state.

Such phenomena include specific heat, critical fields, tunnel-

ling, and others. In addition, predictions by the theory have

stimulated exploratory experimental work which has resulted in

uncovering new phenomena associated with the superconducting

state. A detailed examination of BCS theory will be given in the

next paper (17). Here, only predictions concerning T_c will be

considered.

In BCS theory, the transition to the superconducting state

is given by

$$k_B T_c = (h/2\pi)\, \omega_o \exp[-1/N(0)V] \qquad (\underline{13})$$

where ω_o is the Debye frequency of the lattice, $N(0)$ is the

density of states at the Fermi surface, and V is the effective

electron-electron attraction. McMillan (18) gives Equation (14)

for strongly coupled superconductors

$$T_c = (\theta_D/1.45)\exp[-1.04(1 + \lambda)/(\lambda - \mu^*(1 + \langle\omega\rangle\lambda/\omega_o)] \qquad (\underline{14})$$

where θ_D is the Debye temperature, $\langle\omega\rangle$ is the average phonon

energy, and μ^* is a Coulomb pseudo-potential. The parameter λ is

related to the electronic density of states through the equation

$$\lambda = N(0)\langle I^2\rangle/M\langle\omega^2\rangle \tag{15}$$

where $\langle I^2\rangle$ is an electron-phonon matrix element averaged over the

Fermi surface, M is the atomic mass, $\langle\omega^2\rangle$ is the mean square

phonon frequency.

Equation (13) yields clues concerning enhancement of

superconduction transition temperatures. The Debye frequency may

be obtained from a model in which the positive ions of mass M are

attached to lattice points by springs having a force constant $\underline{k}$.

From Hooke's law

$$\omega_o = (k/M)^{1/2} \tag{16}$$

If the force constant is increased, then it may be seen from

Equation (13) that T_c will increase, but there is a correlation

between the force constant and V, the electron-electron attrac-

tion. An increase in $\underline{k}$ will lead to an increase in the rigidity

of the lattice and, in turn, to a diminished distortion of the

lattice by the electron. In the rigid limit $\underline{V}$ will become very

small and T_c will vanish. A reduction in $\underline{k}$ will lead to a softer

lattice, an increased electron-electron interaction, and initial-

ly an enhancement of T_c. There is a limit to enchancement of T_c

by reducing $\underline{k}$. At some point the decrease in ω_o will outweigh

the benefit gained from the increase in $\underline{V}$.

The Debye frequency can be increased by decreasing $\underline{M}$, the

mass of the positive ions in the lattice. This feature of BCS theory has stimulated interest in the design of systems (19)

which utilize electron-hole pairs or excitons, because of their

small $\underline{M}$'s. Calculations on proposed model systems have led to

predictions of very high temperature superconducting transitions

for systems described by Equation (13). It is unclear at this

time whether the new high temperature ceramic superconductors

exhibit BCS-type superconductivity.

Josephson Effect When a voltage difference is applied between

two superconducting bodies separated by a insulating barrier of

about 10^{-8} m, the unpaired electrons are able to tunnel across

the barrier. The passage of electrons sets up a current which is

characteristic of the magnitude and direction of the voltage

difference. If the insulating barrier is made even thinner, then

Cooper pairs may tunnel from one side of the barrier to the other

side. However, there are important differences in the tunneling

of unpaired electrons and the tunneling of Cooper pairs. The

current from the Cooper pairs flows with zero resistance as in a

single superconductor, and the electrons pass and current flows

in the absence of a voltage difference. The system of two

superconductors and a thin barrier behaves as a single supercon-

ductor. As noted above, all Cooper pairs must have the same

momentum and their density must be uniform. Thus, the Cooper

pairs pass through the barrier in order to achieve a uniform

density and equal momentum in the composite superconducting

system. This behavior was predicted by Josephson (16) when he

was a graduate student at Cambridge University.

The magnitude of the Josephson current is limited. If the current were above a sufficently large value, the Cooper pairs passing through the barrier may have enough energy to dissociate into unpaired electrons. The unpaired electrons do not move with zero resistance, nor do they move in the absence of a voltage difference. These physical constraints limit the magnitude of the Josephson current.

The Josephson effect is very sensitive to the presence of external magnetic fields. Changes of external fields on the order of one part in 10^{10} may be detected. Josephson junctions are widely used in ultrasensitive magnetometers and voltmeters.

5   THE INTERMEDIATE STATE AND THE MIXED STATE

Effects of sample shape on the internal magnetic field of a body were ignored in Equation (6). They must now be taken into consideration. The magnetic field $\underline{H}^i$ within a body is related to the applied magnetic field $\underline{H}$ by the relationship

$$H^i = H - DM \tag{17}$$

where $\underline{D}$ is the demagnetization factor which depends on the shape of the specimen. For the purposes of this discussion, only spherical bodies, for which $\underline{D} = 4\pi/3$, will be considered. Along a specific direction in the magnetic field, $B = H^i + 4\pi M$ in the normal state. In the superconducting state, $\underline{B} = 0$, and for a Type I superconductor $H^i = -4\pi M$. Thus,

$$M = -H/(4\pi - D) \tag{18}$$

With $D = 4\pi/3$ for a spherical body, then

$$M = -3H/8\pi \tag{19}$$

and

$$\chi = -3/8\pi \tag{20}$$

Considering the spherical body, the maximum value of the internal field occurs on the equator, and it is given by

$$H^i = 3H/2 \tag{21}$$

This has an important consequence for superconductivity in that the applied magnetic field will exceed the critical field when

$$H > 2H_c/3 \tag{22}$$

and regions of the sphere must go normal. The entire sphere cannot go normal since $\underline{H} < \underline{H}_c$; the entire sphere can go normal only when $\underline{H} = \underline{H}_c$. Therefore, in the region $2H_c/3 < H < H_c$, both normal and superconducting regions of the material coexist. This state is called the intermediate state. The intermediate state consists of interleaved thin slabs of material in the normal and superconducting states, and it exists only in the presence of an applied magnetic field.

THE MIXED STATE   The physical picture of the mixed state between H_{c1} and H_{c2} is that of a bundle of filaments of normal-state material in flux tubes arranged parallel to the applied flux lines, and these flux tubes are bathed in a sea of superconducting material. A schematic representation of the mixed state is given in Figure 6. Experimental manifestations of the mixed state include a discontinuity in the specific heat at the superconducting state-mixed state transition H_{c1}, and an abrupt

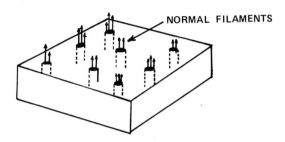

Figure 6. the mixed state of a Type II superconductor. The normal filaments inbeded in a superconducting mass are shown.

change in the specific heat at the mixed state—normal state transition H_{c2}.

The flux through a normal filament is accompanied by an electric current flowing around the filament in a plane perpendicular to the filament, something akin to a whirlpool. It is for this reason that the normal filaments are called <u>vortex lines</u>. The flux through a vortex line is quantized and has values equal to integral multiples of h/2e, the basic quantum of flux.

6 HIGH-TEMPERATURE SUPERCONDUCTING CERAMIC MATERIALS

<u>The Report That Stimulated the Current Interest</u>   On January 27, 1986, Muller and Bednorz (20) observed a strong decrease in resistivity upon cooling a barium-lanthanun-copper-oxide ceramic material. This ceramic, which was reported by Michel and Raveau

(21) in 1984, was selected for study because it exhibits a number
of oxygen-deficient phases with mixed valence copper sites. It
was anticipated that the itinerant electronic states between the
non-Jahn Teller copper(III) and Jahn-Teller copper(II) ions would
have considerable electron-phonon coupling and metallic conductivity.

Precedent Systems Superconductivity had been observed in other
conducting oxides. The compound $Li_{1+x}Ti_{2-x}O_4$ was reported to
have an onset temperature of 13.7 K in 1973 (22), and a T_c of 13
K was reported in 1975 for $BaPb_{0.7}Bi_{0.3}O_3$ (23). Both of these
materials have mixed valence metals (Ti^{3+}/Ti^{2+} and Bi^{5+}/Bi^{3+}),
and superconductivity occurs only over a small range of dopant
concentration, x, with the highest transition temperatures
occurring near metal-insulator phase boundaries.

The Confirming Evidence In the Conclusion of their paper, Bednorz
and Muller state, somewhat tentatively, that "Samples annealed
near 900 C under reducing conditions show features associated
with an onset of granular superconductivity near 30 K". The
onset of superconductivity was confirmed in single phase samples
of $La_{2-x}Ba_xCuO_4$, and the results were announced in late 1986 by
Chu and coworkers in Houston (22) and Tanaka, Kitazawa and
coworkers in Japan (23). Based on powder X-ray diffraction data,
the high temperature superconducting material with x = 0.15 was
reported to be tetragonal and to have the layered K_2NiF_4
structure. The sample was prepared by a solid-state reaction of
a mixture of lanthanum oxide, barium carbonate, and copper oxide

at temperatures above 900 C. The Japanese group found a superconducting onset at 32 K with zero resistance being attained at
22 K, with 30% of the sample showing bulk diamagnetism. The
Houston group found the onset of superconductivity to be 39 K
with full diamagnetism reached at 22 K with 10% of the sample
fully superconducting. The Houston group also noted a sharp
decrease in resistance near 70 K, but the phase responsible for
the behavior was not determined. Both groups noted the critical
temperature was dependent on the annealing conditions of the
pressed pellets. Best results were obtained with pellets
annealed under oxygen and then cooled in an oxygen atmosphere.
Communication of Results Following these announcements, there
began a flurry of activity and many advances were made. Even
rapid communication in scientific journals was slow compared to
the pace of discoveries in high temperature superconductivity.
However, newspapers and news magazines publish rapidly, and much
of the new information was communicated through the pages of the
Wall Street Journal, the New York Times, Science, Chemical and
Engineering News, and others including the evening national news
broadcasts. Sometimes the reports were premature, and some
erronous information was reported.
Studies Under Pressure Samples with the K_2NiF_4 layered structure
were subjected to pressure, and it was observed that the onset
temperature was enhanced at a rate of 1 deg/kbar. This rate is
about 100 times that of conventional superconductors (24). Chu
and co-workers suggested that the large pressure effect might be

due to interfacial superconductivity (26) between a metal and a
semiconductor where pressure can modify the coupling between the
two components. They also discussed the role of the mixed
valence copper ions in the superconducting mechanism.

It apparently occurred to several groups simultaneously to
substitute strontium for barium, and there soon were several
reports of near 40 K superconductivity (27,28). Strontium was
considered promising because its ionic radius is closer to that
of the lanthanum ion than is the radius of barium, and there
could possibly be fine tuning of the copper(III):copper(II) ratio
without distortion of crystal structure. The calcium analogue
was also prepared and characterized (28), where an onset temperature of 23 K was found for single phase $La_{1.85}Ca_{0.15}CuO_4$. On the
basis of the pressure studies by Chu and coworkers and arguments
concerning internal pressure, it is not surprising that the
calcium doped system has a suppressed T_c.

The next logical experimental tack was to replace the
lanthanum ion by other dopants, and early successful syntheses
utilizing yttrium yielded polyphasic material samples with
transition temperatures in excess of 90 K (29). The superconducting phase was identified as a black material with formula
$YBa_2Cu_3O_y$ (where y is now given as [7 - x]) (30). The timeline
for research developments on high temperature superconductors is
given later in these proceedings by Painter, et al.

Processing and Applications   For superconducting materials to be
useful in devices they must carry substantial electrical currents

and they must withstand relatively large applied magnetic fields. In May, 1987 workers at IBM's Thomas J. Watson Research Laboratory prepared a thin-film crystalline specimen of the high T_C material and found that it could conduct 100,000 amperes per square centimeter at 77 K. The goal is 1,000,000 amperes per square centimeter. This result points to applications of the these new materials, for example in SQUIDS, as demonstrated by IBM and the National Bureau of Standards.

Much effort has been expendend in attempts to make wires out of the ceramic materials. The approach at AT&T Bell Laboratories was to place the powdered oxide in a thin metal tube, which was then stretched, shaped, and fired to yield a superconducting "wire", while at Argonne the powder was mixed with a binder, shaped, and fired. The binder burned away during the firing process, and the superconducting wire was left as shaped.

There are numerous applications for high temperature superconducting materials. Superfast computers could be produced using Josephenson junctions which are more than 10 times as fast as conventional switches. Small, but more powerful magnets could be produced from superconducting wires, and these could be used in numerous transportable devices currently not possible with heavier conventional systems. Frictonless devices, including levitated trains, may also be envisaged. A detailed considera-tion of applications of high temperature superconductors is presented by Painter in Chapter 3.

ACKNOWLEDGMENT   This work was supported in part by the Office of Naval Research.

REFERENCES

1. H. Kammerlingh Onnes, Leiden Comm., 120b, 122b, 124c (1911).

2. B. W. Roberts, Handbook of Chemistry and Physics, 61st Ed., E-87, 1980.

3. B. W. Roberts, NBS Technical Note 482 (1969); NBS Technical Note 983 (1978).

4. (a) J. R. Gavaler, Appl. Phys. Lett., 23, 480 (1973).

    (b) L. Testardi, J. H. Wermick, W. A. Roger, Solid State Commun., 15, 1 (1974).

5. J. Bardeen, L. N. Cooper, and J. R. Schrieffer, Phys. Rev., 108, 1175 (1957).

6. (a) M. A. Biondi, M. P. Garfunkel, and A. O. McCoubrey, Phys. Rev., 101, 1427 (1956).

    (b) R. E. Glover and M. Tinkham, Phys. Rev., 104, 844 (1956).

    (c) R. E. Glover and M. Tinkham, Phys. Rev., 108, 243 (1957).

7. W. Meissner and R. Ochsenfeld, Naturwissenschaften, 21, 787 (1933).

8. F. London and H. London, Proc. Roy. Soc. (London), A149, 72 (1935); Physica, 2, 341 (1935).

9. A. B. Pippard, Proc. Roy. Soc. (London), A216, 547 (1953).

10. T. E. Faber and A. B. Pippard, Proc. Roy. Soc. (London), A231, 336 (1955).

11. V. L. Ginsberg and L. D. Landau, Zh. Eksperim. i Teor. Fiz., 20, 1064 (1950).

12. A. A. Abrikosov, Zh. Eksperim. i Teor. Fiz., 32, 1442 (1957).

13. L. N. Cooper, Phys. Rev., 104, 1189 (1956).

14. H. Frohlich, Phys. Rev., 79, 845 (1950).

15. (a) E. Maxwell, Phys. Rev., 78, 477 (1950).

    (b) C. A. Reynolds, B. Serin, W. H. Wright, and L. B. Nesbitt, Phys. Rev., 78, 487 (1950).

16. B. D. Josephson, Phys. Lett., 1, 25 (1962).

17. J. H. Miller, Jr., following paper.

18. W. L. McMillan, Phys. Rev., 167, 331 (1968).

19. W. A. Little, Intl. J. Quantum Chem.: Quantum Chem. Symposium, 15, 545 (1981).

20. J. G. Bednorz and K. A. Muller, Z. Phys. B - Condensed Matter, 64, 189 (1986).

21. C. Michel and B. Raveau, Chim. Min., 21, 407 (1984).

22. D. C. Johnston, H. Prakash, W. H. Zachariasen, and R. Viswanathan, Mat. Res. Bull., 8, 777 (1973).

23. A. W. Sleight, J. L. Gillson, and F. B. Bierstedt, Solid State Commun., 17, 27 (1975).

24. C. W. Chu, P. H. Hor, R. L. Meng, L. Gao, Z. J. Huang, and Y. Q. Wang, Phys. Rev. Lett., 58, 405 (1987).

25. (a) S. Uchida, H. Takagi, K. Kitazawa, and S. Tanaka, Jpn. J. Appl. Phys., 26, L1 (1987).

(b) H. Takagi, S. Uchida, K. Kitazawa, and S. Tanaka, Jpn. J. Appl. Phys., 26, L123 (1987).

26. D. Allander, J. Bray, and J. Bardeen, Phys. Rev. B, 3, 1020 (1973).

27. (a) Z. X. Zhao, L. Q. Chen, C. G. Cui, Y. Z. Huang, J. X. Liu, G. H. Chen, S. L. Li, S. Q Guo, and Y. Y. He, Kexue Tongbao, 32, 522 (1987).

     (b) R. J. Cava, R. B. van Dover, B. Batlogg, and E. A. Rietman, Phys. Rev. Lett., 58, 408 (1987).

     (c) J. M. Tarascon, L. H. Greene, W. R. McKinnon, G. W. Hull, and T. H. Geballe, Science, 235, 1373 (1987).

     (d) J. G. Bednorz, K. A. Müller, and M. Takashige, Science, 236, 73 (1987).

28. H. Takagi, S. Uchida, K. Fuchi, and S. Tanaka, Chem. Lett., 429 (1987).

29. (a) M. K. Wu, J. R. Ashburn, C. J. Rorng, P. H. Hor, R. L. Meng, L. Gao, Z. J. Huang, Q. Wang, and C. W. Chu, Phys. Rev. Lett., 58, 908 (1987).

     (b) C. W. Chu, P. H. Hor, R. L. Meng, L. Gao, Z. J. Huang, Y. Q. Wang, M. K. Wu, J. R. Ashburn, and C. Y. Huang, Phys. Rev. Lett., 58, 911 (1987).

     (c) Z. Ahao, C. Liquan, y. Qiansheng, H. Yuzhen, C. Genghua, T. Ruming, L. Guirong, C. Changgeng, C. Lie, W. Lianzhong, G. Shuquan, L. Shanlin, and B. Jianqing, Kexue Tongbao, No. 6, (1987).

(d) P. Ganguly, A. K. Raychaudhuri, K. Sreedhar, and C. N. R. Rao, Pramana- J. Phys., 27, L229 (1987).

30. (a) P. M. Grant, R. B. Beyers, E. M. Engler, G. Lim, S. S. P. Parkin, M. L. Ramirez, V. Y. Lee, A. Nazzal, J. E. Vazquez, and R. J. Savoy, Phys. Rev. B, 35, 7242 (1987).

(b) J. M. Tarascon, L. H. Greene, W. R. McKinnon, and G. W. Hull, Phys. Rev. B, 35, 7115 (1987).

(c) R. J. Cava, B. Batlogg, R. B. van Dover, D. W. Murphey, S. Sunshine, T. Siegrist, J. R. Remeika, E. A. Rietman, S. Zahurak, and G. P. Espinosa, Phys. Rev. Lett., 58, 1676 (1987).

3

High-Temperature Superconduction: An Overview

WID J. PAINTER
Guilford Technical Community College, Jamestown, NC 27282

SECTION 1: INTRODUCTION

1.1 Ba-La-Cu-O Superconductor

In 1973, the highest critical temperature of superconductors was
only 23K in alloys of niobium and germanium. Thirteen years later,
Alex Müller and Georg Bednorz at the IBM Zurich Research Laboratory
in Switzerland successfully synthesized a new type of superconducting
material which consisted of a ceramic oxide of barium, lanthanum,
and copper (Ba-La-Cu-O). They observed a dramatic drop in resistivity
at 35K, and recognized the possibility of this type of material being
a high temperature superconductor.

Two months after the publication of Müller and Bednorz's
discovery, Shoji Tanaka at the University of Tokyo and Ching-Wu Paul
Chu at the University of Houston each independently confirmed the
work done in Switzerland.

In December of 1986, Tanaka's group determined the structure of
the new ceramic superconductor, and found its molecular formula to
be $La_{1.85}Ba_{0.15}CuO_4$. Several other research groups studied the
Ba-La-Cu-O superconductor, including Jorgensen, Schreitter, et. al.,
at Argonne National Laboratories and C. W. Chu, et. al., at the
University of Houston.

67

1.2 Sr-La-Cu-O Superconductor

Early in 1987, three groups working independently found that the
critical temperature of the new ceramic superconductor could be
raised five degrees to 40K by replacing the barium atoms with
strontium. The three groups were led by Tanaka at the University
of Tokyo, Chu at the University of Houston-Huntsville, and several
researchers at AT & T Bell Laboratories-Murray Hill, including
Van Dover, Cava, Batlogg, and Rietman.

Other groups have verified and characterized the Sr-La-Cu-O
superconductor, including Wang, Geiser, et. al., at Argonne National
Laboratory, Gangula and Rao at the Indian Institute of Science,
Tanaka at the University of Tokyo, and Politis, Geerk, et. al.,
at the Institute for Nuclear Solid State Physics in Karlsruhe,
West Germany.

1.3 Y-Ba-Cu-O Superconductor

The jewel in the crown of ceramic superconductors, however, was
published in March of 1987 when Chu and Wu's groups synthesized
$YBa_2Cu_3O_7$, which had a T_c of an incredible 90 - 100 K.

Within a few days, several groups were repeating the synthesis
and characterizing the new high temperature superconductor,
including teams led by Stacy at UC-Berkeley, and Engler and Grant
at IBM-Almaden. In addition, the new material's structure was soon
shown to be similar to a type of mineral called perovskite by several
groups already working in the field, including those at Bell Laboratories, IBM-Almaden, and Argonne National Laboratory. Kaiser and
Holtzberg of IBM grew single crystals of $YBa_2Cu_3O_7$ with dimensions
in the millimeter range.

Several researchers have observed unusual drops in resistance
as high as 240K ($-33^{\circ}C$), but the general consensus at the present
time is that these are not stable superconductors. Results have
not been reproducible, and are believed to be due to factors such
as resistance changes at the contact leads or perhaps extreme
quantities of microscopic heterogeniety in the sample.

1.4 Ho–Ba–Cu–O Superconductor

Leigh, Porter, et. al., at Argonne National Laboratory synthesized
$HoBa_2Cu_3\delta_{6+x}$ and found a critical temperature of 93.5K.

1.5 Superconductivity Time Line

Figures 1 and 2 show the historical development of superconduction.
Note the virtual explosion of papers in 1987 following the discovery
of the Ba–La–Cu–O superconducting ceramic by Müller and Bednorz
in 1986.

SECTION 2: METHODS OF SYNTHESIS

Currently, there are two general methods of synthesizing the
high temperature superconductors: the solid state fusion method,
and the solutions method.

2.1 Solid state Fusion Method

The most popular method of the two is the solid state fusion technique. In this procedure, a finely ground mixture of the oxide or
carbonate powders of the desired elements is heated for 12 hours
at temperatures ranging between 900 and $1100^{o}C$, sometimes in an
atmosphere of oxygen. The resulting powder is reground and heated
as previously described. The product is subsequently cold–pressed
into pellets at 1500 psi and heated at 700 – $900^{o}C$ for several
hours and slowly cooled to room temperature over a period of a
few hours.
      The solid state fusion method has been used by Engler, Lee,
et. al., at IBM–Almaden to synthesize a variety of superconducting
ceramics including Y–Ba–Cu–O, Nd–Ba–Cu–O, Sm–Ba–Cu–O, Eu–Ba–Cu–O,
Gd–Ba–Cu–O, Dy–Ba–Cu–O, Ho–Ba–Cu–O, Yb–Ba–Cu–O, Lu–Ba–Cu–O,
Y–Sc–Ba–Cu–O, Y–La–Cu–O, Y–Lu–Cu–O, Y–Sr–Ca–Cu–O, Y–Ba–Sr–Cu–O,
Y–Ba–Ca–Cu–O, Yb–Ba–Sr–Cu–O, and Yb–Ba–Ca–Cu–O. All of them except

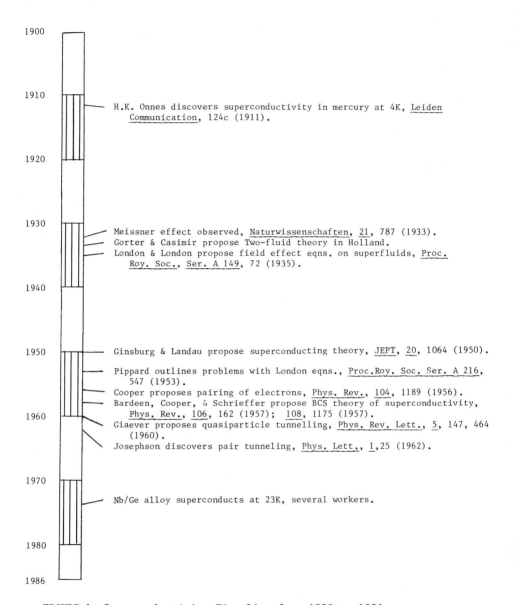

FIGURE 1   Superconductivity Time Line from 1900 to 1986.

1986

Bednorz & Muller make Ba-La-Cu-O Superconductor at 35K,
Z. Phys., B64, 189 (1986).

1987

Others continue work on Ba-La-Cu-O superconductors:
Uchida, Takagi, Kitazawa, Tanaka, Jpn. J. Appl. Phys., Part 2, 26,
    L151 (1987).
Takagi, Uchida, Kitazawa, Tanaka, Jpn. J. Appl. Phys., Part 2, 26,
    L123 (1987).
Jorgensen, Schuettler, Hinks, Capone, Zhang, Brodsky, Scalapino,
    Phys. Rev. Lett., 58(10), 1024 (1987).
Bednorz, Takashige, Muller, Europhys. Lett., 3(3), 379 (1987).
Chu, Hor, Meng, Gao, Huang, Science, 235(4788), 567 (1987).

Several investigators substitute Ca & Sr for Ba:
Uchida, Takagi, Kitazawa, Tanaka, Jpn. J. Appl. Phys. Lett., 26,
    L1-L2 (1987).
Ganguly & Rao, Proc. Indian Acad. Sci., Chem. Sci., 97, 631 (1987).
Van Dover, Cava, Batlogg, Rietman, Phys. Rev. B, 35(10), 5337 (1987).
Kanbe, Kishio, Kitazawa, Fueki, Takagi, Tanaka, Chem. Lett., 3,
    547 (1987).
Bednorz, Muller, & Takashige, Science, 236(4797),. 73 (1987).
Tanaka, Jpn. J. Appl. Phys., Part 2, 26(3), L203 (1987).
Tanaka, et. al., Jpn. J. Appl. Phys., Part 2, 26(3), L196 (1987).
Politis, Geerk, Dietrich, Obst, Z. Phys. B, 66(2), 141 (1987).
Wang, Geiser, Thorn, Douglas, Beno, Monaghan, Allen, Proksch,
    Stupka, et. al., Inorg. Chem., 26(8), 1190 (1987).
Cava, van Dover, Batlogg, Reitmann, Phys. Lett., 58(4), 408 (1987).

Several workers investigate Y-Ba-Cu-O superconductors:
Bourne, Cohen, Creager, Crommie, Stacy, & Zettl, Phys. Lett. A,120,
    494 (1987).
Engler, Lee, Nazzal, Beyers, Lim, Grant, Parkin, Ramirez, Vazquez,
    & Savoy, J. Am. Chem. Soc., 109, 2848 (1987).
Stacy, Badding, Geselbracht, Ham, Holland, Hoskins, Keller,
    Millikan, Zur Loye, J. Am. Chem. Soc., 109, 2528 (1987).

A group worked with (Ba-Sr)-La-(Hg-Ag)-Cu-O compounds:
Saito, Noji, Endo, Matsuzaki, Katsumata, Jpn. J. Appl. Phys.,
    Part 2, 26, L223 (1987).

Another group synthesized & characterized Ho-Ba-Cu-O superconductor:
Leigh, Porter, Thorn, Geiser, Umezawa, Wang, Kwok, Kao, Monaghan,
    Crabtree, Carlson, Williams, Inorg. Chem., 26, 1645 (1987).

Organic superconductivity was investigated:
Tachibana, Inoue, Fueno, Yamabe, & Hori, Synth. Met., 19, 99 (1987).

FIGURE 2   Superconductivity Time Line for 1986 - 1987.

the Lu–Ba–Cu–O ceramic have critical temperatures in the range
between 80 and 100K.

2.2 Solution Method

In this method which was pioneered by Wang, Geiser, et. al.,
at Argonne National Laboratories, a soluble nitrate salt of each
metal is dissolved in water and neutralized with a KOH solution.
The hydroxyl carbonate salts are quantitatively precipitated with a
potassium carbonate solution. After drying overnight and heating
for three hours, the oxide is quantitatively produced as a black
powder. The oxide is then sintered at $1100^{o}C$ for two hours,
resulting in the superconducting pellet of the material.

The Argonne group of Wang, Geiser, et. al., used their method
to produce the La–Sr–Cu–O superconductor, and found that a single
crystal has a critical temperature of 8.5 to 9.0K.

SECTION 3: FABRICATION METHODS

The immediate short-term goal of the federal superconductivity
effort is concentrated on two major goals: (1) developing a
piece of superconducting wire, and (2) producing a strip of
superconducting film.

The current status of these objectives was outlined in an
article published in the May 11, 1987 issue of Chemical and Engineering News on pages 14 through 16, and is summarized below.

3.1 Wire Fabrication

Some of the methods to produce a wire of superconducting ceramic
include: (a) filling a thin, hollow tube with Y–Ba–Cu–O powder
and drawing the tube out to the desired diameter; (b) mixing the
powder with an organic polymer, extruding it into a thin wire,

and burning off the polymer to produce the superconducting wire;
and (c) coating the superconducting powder onto a metal wire, braiding
the composite into bundles, and saturating them with copper to form
strands of wire able to carry useful electrical currents.

3.2 Fabrication of Strips

Thin films of superconducting Y-Ba-Cu-O ceramic composite have
been fabricated by using an evaporation/condensation process. In
another technique, the ceramic powder is mixed with an organic
polymer and extruded into a thin, flexible tape. The organic
component is then burned off leaving a thin layer of the superconducting material.

SECTION 4: APPLICATIONS

Superconducting applications can be divided into two major categories:
small-magnitude devices and large-magnitude devices.

4.1 Small-magnitude Devices

There are already several of these devices in use, such as superconducting wire formed with a diameter of 0.01 inches so as to
remain flexible.

Another example of devices already in use are a group of
solid state devices known as Josephson junctions. This superconducting element can switch as fast as one picosecond, which is
nearly ten times faster that switches previously used.

Superconducting film will be able to conduct electricity
without generating heat, so components can be located more compactly
on a circuitboard, thereby increasing the speed of the device.
This will result in smaller, more powerful supercomputers.

4.2 Large-magnitude Devices

Probably the most exciting application of large-scale superconductivity is the $4.5 billion superconducting super collider
(SSC), which will be the world's largest high energy accelerator.
The SSC will rotate beams of protons to energies as high as
20 trillion electron volts, and will lead to new insights
regarding the nature of matter. Superconducting coils will
create intense magnetic fields in the 10,000 superconducting
magnets contained in the device.

Another area of application of superconductivity is in the
large-scale generation, storage, and transmission of electrical
power. Superconducting wire forming generator coils without any
wasteful heat could generate electrical current. This current, in
turn, could be stored in large, superconducting coils without
any loss due to resistance. Finally, superconductive transmission
lines could carry the current for thousands of miles without
any loss of current.

A final application of large-scale superconductivity is
seen in trains levitated by superconductive magnets. These trains
could travel at hundreds of miles per hour hovering a mere four
inches above the track without the hindrance of friction.

SECTION 5: BIBLIOGRAPHY

5.1 Superconductivity before 1986

1. H. K. Onnes, Leiden Communication, 124c (1911).

2. W. Meissner and R. Ochsenfeld, Naturwissenschaften, 21, 787 (1933).

3. F. London and H. London, Proc. Roy. Soc., Ser. A149, 72 (1935).

4. V. L. Ginsburg and L. D. Landau, JEPT, 20, 1064 (1950).

5. A. B. Pippard, Proc. Roy. Soc., Ser. A216, 547 (1956).

6. L. N. Cooper, Phys. Rev., 106, 1189 (1956).

7. J. Bardeen, L. N. Cooper, and J. R. Schrieffer, Phys. Rev., 106,
        162 (1957); 108, 1175 (1957).

8. I. Giaever, Phys. Rev. Lett., 5, 147 (1960).

9. B. D. Josephson, Phys. Rev. Lett., 1, 251 (1962); Adv. Phys., 14, 419 (1965).

10. J. R. Gavaler, Appl. Phys. Lett., 23, 480 (1973).

5.2 Ba-La-Cu-O Superconductor(s)

1. J. G. Bednorz and K. A. Müller, Z. Phys., B64, 189 (1986).

2. S. Uchida, H. Takagi, K. Kitazawa, and S. Tanaka, Jpn. J. Appl., Phys., Part 2, 26, L151 (1987).

3. H. Takagi, S. Uchida, K. Shinichi, and S. Tanaka, Jpn. J. Appl. Phys., Part 2, 26, L123 (1987).

4. J. D. Jorgensen, H.B. Scheuttler, D. G. Hinks, D. W. Capone II, K. Zhang, M. B. Brodsky, and D. J. Scalapino, Phys. Rev. Lett., 58, (10), 1024 (1987).

5. J. G. Bednorz, M. Takashige, and K. A. Müller, Europhys. Lett., 3 (3), 379 (1987).

6. C. W. Chu, P. H. Hor, R. L. Meng, L. Gao, and Z. J. Huang, Science, 235 (4788), 567 (1987).

5.3 Ca/Sr-La-Cu-O Superconductor(s)

1. S. Uchida, H. Takagi, K. Kitazawa, and S. Tanaka, Jpn. J. Appl. Phys. Lett., 26, L1-L2 (1987).

2. P. Ganguly and C.N.R. Rao, Proc. Indian Acad. Sci., Chem. Sci., 97, 631 (1987).

3. R. B. Van Dover, R. J. Cava, B. Batlogg, and E. A. Rietman, Phys. Rev. B, 35 (10), 5337 (1987).

4. S. Kanbe, K. Kishio, K. Kitazawa, K. Fueki, H. Takagi, and S. Tanaka, Chem. Lett., 3, 547 (1987).

5. J. G. Bednorz, K. A. Müller, and M. Takashige, Science, 236 (4797), 73 (1987).

6. S. Tanaka, Jpn. J. Appl. Phys., Part 2, 26 (3), L203 (1987).

7. S. Uchida, H. Takagi, S Tanaka, K. Nakao, N. Miura, K. Kishio, K. Katazawa, and K. Fueki, Jpn. J. Appl. Phys., Part 2, 26 (3), L196 (1987).

8. Hau H. Wang, U. Geiser, R. J. Thorn, Douglas, M. A. Beno, M. R.
    Monaghan, T. J. Allen, R. B. Proksch, D. L. Stupka, et. al.,
    Inorg. Chem., 26 (8), 1190 (1987).

9. R. J. Cava, R. B. Van Dover, B. Batlogg, and E. A. Rietman,
    Phys. Lett., 58 (4), 408 (1987).

5.4 Y-Ba-Cu-O Superconductor(s)

1. M. K. Wu, J. R. Ashburn, C. J. Torng, P. H. Hor, R. L. Meng, L.
    Gao, Z. J. Huang, Y. Q. Wang, and C. W. Chu, Phys. Rev. Lett.,
    58, 908 (1987).

2. L. C. Bourne, M. L. Cohen, W. N. Creager, M. F. Crommie, A. M.
    Stacy, and A. Zettl, Phys. Lett. A, 120 494 (1987).

3, E. M. Engler, V. Y. Lee, A. I. Nazzal, R. B. Beyers, G. Lim,
    P. M. Grant, S. S. P. Parkin, M. L. Ramirez, J. E. Valquez, and
    R. J. Savoy, J. Am. Chem. Soc., 109, 2848 (1987).

4. A. M. Stacy, J. V. Badding, M. J. Geselbracht, W. K. Ham, G. F.
    Holland, R. L. Hoskins, S. W. Keller, C. F. Millikan, and H. C.
    Zur Loy, J. Am. Chem. Soc., 109, 2528 (1987).

5.5 Ba/Sr-La-Hg/Ag-Cu-O Superconductor(s)

1. Y. Saito, T. Noji, A. Endo, N. Matsuzaki, and M. Katsumata, Jpn.
    J. Appl. Phys., Part 2, 26, L223 (1987).

5.6 Ho-Ba-Cu-O Superconductor(s)

1. C. Leigh, C. Porter, R. J. Thorn, U. Geiser, A. Umezawa, M. R.
    Managhan, et. al., Inorg. Chem., 26, 1645 (1987).

5.7 Organic Superconductors

1. A. Tachibana, T. Inoue, H. Fueno, T. Yamabe, and K. Hori, Synth.
    Met., 19, 99 (1987).

5.8   General Review of High-Temperature Superconductivity

1.   R.   Dagani, <u>Chem. Engr. News</u>, <u>65</u>, (19), 7 (1987).

4

The Physics of High-Temperature Superconductivity

JOHN H. MILLER, JR.   Department of Physics and Astronomy, University of North Carolina, Chapel Hill, North Carolina 27599-3255

I. INTRODUCTION

A determination of the microscopic origin of high temperature superconductivity is one of the paramount problems of condensed matter physics and solid state chemistry. This chapter will attempt to present an overview of what is now known about high T_c superconductivity, within the context of some of the proposed mechanisms.

II. EVIDENCE FOR PAIRING

There is ample experimental evidence that the remarkable properties of the new metal-oxide superconductors[1-7] result from a pairing of carriers (electrons or holes), as in the case of conventional superconductors. For example, the Shapiro steps in Josephson tunnel junctions biased with both ac and dc voltages have a spacing $\Delta V = 2e/h\nu$,[8] clearly indicating an effective charge of 2e. Furthermore, the flux quantum in a high T_c superconducting ring has been measured to be $\Phi_0 = h/2e$,[9] corroborating the Shapiro step experiments. All viable theories thus propose a form of carrier pairing as being responsible for high T_c superconductivity. However, at present

(1987) there is little or no consensus on the pairing mechanism-
either on the nature of the virtual excitations which mediate the
pairing or on the type of pairing (BCS vs. Bose-type pairing).
Important issues which must be resolved include whether the Cooper
pairs have s-wave symmetry and the strength of the pairing
interaction.

III. SUMMARY OF PROPOSED MECHANISMS

In principle, a sufficiently strong attractive interaction between
the carriers would allow a Bose condensation of tightly bound pairs,
as originally proposed by Schafroth, Blatt, and Butler.[10] In a
conventional BCS-type superconductor, on the other hand, there are
~10^6 electrons within the average distance between mates of a single
pair,[11] so that the pairs do not obey true Bose-Einstein statistics,
but the Pauli principle plays an important role. This suggests a
classification of theories into   (i) BCS pairing,[12-23] where the
pairs form at the transition temperature and (ii) Bose condensation
of tightly bound pairs at T_c,[24-26] where the pairs already exist
above the transition temperature.   Table I illustrates a
classification scheme proposed by Varma,[12] where the pairing is
mediated by (i) electron-phonon interactions,[13,24] (ii) magnetic
correlation of the electrons (spin fluctuations),[14-18,25,26] and
(iii) exchange of virtual electronic polarization resonances,[19-23]
including excitons,[19-21] charge-transfer excitations,[22] and soft
plasmons.[23]

IV. EVIDENCE AGAINST PHONON-MEDIATED PAIRING

When the transition temperatures of known superconductors are plotted
against the Sommerfeld constant γ, reflecting the density of states
$N(0)$ at the Fermi surface, the superconducting materials appear to
separate into at least three distinct families, as shown in Fig. 1.

TABLE I. Some of the proposed pairing mechanisms for high T_c
superconductivity.

	TYPE OF PAIRING	
	BCS PAIRING	BOSE CONDENSATION
VIRTUAL EXCITATIONS:		
PHONONS	Conventional[13]	Bipolaron formation[24]
MAGNETIC CORRELATIONS	Exchange of antiferromagnetic (AFM) spin fluctuations[14-18]	Resonating valence bond (RVB) or spin-bipolaron formation and Bose condensation[25,26]
ELECTRONIC POLARIZATION RESONANCES	Excitons, electron-hole pairs,[19-21] charge-transfer excitations,[22] soft plasmons[23]	

The heavy Fermion superconductors have anomalously low T_c's vs. γ as compared with "conventional" superconductors, whereas the new oxide superconductors have T_c's which are considerably higher than any previous superconductor with comparable γ, and thus appear to comprise a new family of superconductors as pointed out by Batlogg.[27]

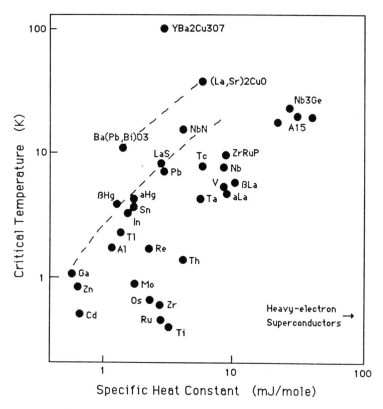

Figure 1. Critical temperature (T_c) *versus* Sommerfeld constant γ, reflecting the density of states at the Fermi surface (ref. 27).

Extensive measurements of the isotope effect[28-33] in $BaPb_{1-x}Bi_xO_3$, $La_{2-x}Sr_xCuO_4$ (LSCO), and $YBa_2Cu_3O_{7-\delta}$ (YBCO) demonstrate that, for a given compound, T_c scales as $M^{-\alpha}$, where M is the mass of the oxygen isotope and α ranges from about zero in the "123" compounds, with T_c's of 90-100 K, to α ~ 0.22 in $BaPb_{1-x}Bi_xO_3$, with a T_c of ~13 K. This indicates that the role of the phonons becomes relatively less important for the higher T_c materials, suggesting that an additional mechanism besides phonon-mediated pairing is largely responsible for superconductivity in the high-T_c "123" compounds. Recently, Gurvitch and Fiory (GF)[34] have reported on the temperature-dependent resistivities of LSCO and YBCO over the temperature range from T_c to 1100 K. If the electron-phonon interactions were strong, the resistivities would begin to saturate at high temperatures, as is observed in V_3Si. However, the measured resisitivies of the new oxide superconductors do *not* saturate at high temperatures, and the GF results place upper bounds on McMillan's[35] electron-phonon coupling constant of $\lambda \leq 0.1$ and $\lambda \leq 0.3$ for LSCO and YBCO, respectively. Such small values of the electron-phonon coupling constant λ provide convincing evidence that phonon-mediated pairing cannot be the sole mechanism responsible for high T_c superconductivity.

V. BCS-TYPE THEORIES

An important unresolved issue concerns why the T_c's of the new copper-oxide superconductors are so much higher than those of conventional superconductors. According to the BCS theory,[36] in the weak-coupling limit the transition temperature is given by

$$k_B T_c = 1.14 \ \hbar\omega_c \exp\left(\frac{-1}{N(0)V}\right) \qquad (1)$$

where V represents an average attractive electron-electron interaction, $N(0)$ is the density of states at the Fermi surface, and $\hbar\omega_c$ is a cut-off energy which represents the energy range, measured from the Fermi surface, over which the electron-electron interaction is attractive. This cut-off energy is comparable to the maximum energy of the excitations which mediate the pairing. For the case of phonon-mediated pairing, ω_c is comparable to the Debye frequency ω_D. Within the BCS framework, there are three ways to increase T_c: (1) increase $N(0)$, (2) increase V, and (3) increase ω_c. In the strong coupling limit, Eq. (1) no longer applies, but qualitatively the same trends still hold.[37] A complication is that V and ω_c are not independent for a given pairing mechanism, but are closely related. An increase in V, by strengthening the electron-phonon interaction for example, often occurs at the expense of a reduction in ω_c.

Since it is now known that $N(0)$ is rather low for the high T_c oxide superconductors, the only possiblities, within the BCS framework, are an enhanced attractive interaction V and/or an enhanced cut-off frequency ω_c. The GF experimental results make it highly unlikely that the observed high T_c's simply result from a large attractive interaction between carriers. It thus appears that the fraction of conducting carriers which participate in the pairing process must be greatly enhanced for the high T_c oxide superconductors, as compared with conventional superconductors.

According to the early theories of high T_c superconductivity proposed by Little,[19] Ginsburg,[20] and Allender et al,[21] an attractive interaction between carriers can be mediated by the exchange of virtual electron-hole pairs- the so-called "excitonic" mechanism for superconductivity. In order for this mechanism to work two distinct types of carriers are required- conducting carriers in a metallic region (chain or plane) which pair to form the superconducting condensate; and polarizable electrons in adjacent region (side chain or plane) which interact with the metallic carriers to mediate the pairing process. It is possible that the polarizable, nonconducting electrons in the Ba-O planes or the carriers in the Cu-O chains may mediate a pairing between the conducting carriers in the Cu-O planes in the "123" compounds. On the other hand, according to the theory of charge transfer excitation exchange, proposed by Varma et al,[12,22] the virtual electronic polarization occurs through the transfer of carriers between the copper and oxygen sites within a Cu-O plane. In this picture the same carriers which mediate the pairing process also participate in forming the superconducting condensate.

The exciton mechanism is predicted to yield a significantly higher T_c than the phonon mechanism because the cut-off energy $\hbar\omega_c$ is expected to be comparable to electronic excitation energies, which are generally much higher than typical Debye energies. Alternatively, one may use the arguments of the isotope effect to arrive at the same conclusion. In the proposed excitonic system the interaction is mediated by the movement of electrons rather than by the much heavier ions of a phonon superconductor. The transition temperature for an excitonic superconductor would thus be scaled up from that of conventional superconductors by a factor of order $(M_{ion}/m_e)^{\frac{1}{2}}$, if all other parameters were kept constant. Similar arguments hold for theories proposing the exchange of soft plasmons[23] or virtual magnetic excitations, such as antiferromagnetic spin fluctuations.[14-18] Again, the higher energies of the relevant virtual excitations increase the energy range over which the electron-electron interaction is attractive, leading to a higher fraction of paired carriers, and thus a higher T_c.

VI. SUMMARY OF EXPERIMENTAL RESULTS AND THEORETICAL INTERPRETATIONS

Recent positron annihilation experiments on YBCO,[38] where the annihilation rates of injected positrons trapped by defects are measured, show a dramatic reduction in the average positron lifetime as the temperature is reduced below the critical temperature. An important conclusion inferred from these studies is that the electronic density near the oxygen vacancies is higher in the superconducting state than in the normal state, implying that there is an elecronic structure change below T_c. An additional implication is that many, or all, of the conducting carriers in the Fermi sea may be participating in the pairing process, rather than just a narrow band of states near the Fermi surface. W. Y. Ching et al[39] have interpreted the positron lifetime results in terms of an excitonic enhanced superconducting mechanism, where excess electrons are available in the superconducting composite state below T_c, but no such electrons exist above T_c.

In the absence of a consensus on the microscopic origin of high T_c superconductivity, a knowledge of the phenomenological Fermi liquid and Ginzgurg-Landau parameters can help guide in the development (or selection) of a suitable microscopic theory. Bardeen et al[40] have compiled normal and superconducting parameters of YBCO using estimates based on existing experimental data,[7,41] as indicated in Table II. The G-L and Pippard coherence lengths for YBCO, $\xi_{GL} \sim \xi_0 \sim 14$ Å, are considerably shorter than those typically found in conventional superconductors. Since the Pippard coherence length ξ_0 is an approximate measure of the average distance between mates of a Cooper pair, when ξ_0 is sufficiently short, a description in terms of real-space pairing and Bose-Einstein condensation becomes viable, unlike the case of conventional superconductivity. On the other hand, the effective penetration length $\lambda_{eff} \sim \lambda_L \sim 1400$ Å is comparable to the penetration length of an ordinary superconductor. The G-L parameter $\kappa = \lambda_{eff}/\xi \sim 100$ is the largest value for any known superconductor. The "123" REBCO (rare earth-$Ba_2Cu_3O_{7-\delta}$) superconductors are thus the most strongly type-II superconductors known. Most importantly, the copper-oxide superconductors are *intrinsically* type-II. In other words they would remain type-II even in the clean limit, where the mean free path length is much longer than the Pippard coherence length, $\ell \gg \xi_0$.

An important practical implication is that the upper critical field $H_{c_2}(T=0) = \sqrt{2}\kappa H_c(0) \sim 200$ T is about an order of magnitude higher than the highest magnetic fields which can be generated in a typical research laboratory. However, the field-dependent critical current density $J_c(H)$ is observed to drop rapidly with increasing applied magnetic field, at a given temperature.[42] This is because, in type-II superconductors, $J_c(H>0)$ is determined by the depinning of magnetic flux vortices (fluxons) due to their Lorentz interaction with the supercurrent. When the current density exceeds the critical value $J_c(H)$, which is much lower than the theoretical "pair-breaking"

TABLE II. Normal state and superconducting properties of $YBa_2Cu_3O_{7-\delta}$
(from ref. 40).

(a) Normal State Properties: $YBa_2Cu_3O_{7-\delta}$

Carrier density:	$n = k_F^3/(3\pi^2) = 9 \times 10^{21}/cm^3$
Fermi wave vector:	$k_F = 6.5 \times 10^7 cm^{-1}$
Density of states:	$N(0) = m^* k_F/(2\pi^2 \hbar^2) = 3 \times 10^{34}/cm^3 erg$
(per spin)	
Sommerfeld const:	$\gamma = 2\pi^2 k_B^2 N(0)/3 = 3 \times 10^3 erg/cm^3 K^2$
Effective mass ratio:	$m^*/m = 9$
Resistivity (90K):	$\rho = 200\mu\Omega cm$
Mean free path (90K):	$\ell = \hbar k_F/(ne^2\rho) = 17$ Å

(b) Ginzburg-Landau Parameters: $YBa_2Cu_3O_{7-\delta}$

Transition temperature:	$T_c = 90$ K
GL penetration depth:	$\lambda_{GL} = 1400$ Å
GL coherence distance:	$\xi_{GL} = 14$ Å
Ratio:	$\kappa \equiv \lambda_{GL}/\xi_{GL} = 100$
Critical fields	$H_c(0) = 12kOe$
$H_{c1} = H_c(0)\ln\kappa/(\sqrt{2}\,\kappa)$	400 Oe
dH_{c1}/dT	-7 Oe/K
$H_{c2} = \sqrt{2}\,\kappa H_c$	80 – 320 T

current density relevant to type-I superconductors, the flux vortices start to move and thus allow energy to be dissipated. By artificially introducing impurities which pin the flux vortices, hopefully without adversely affecting the superconducting properties, it should be possible to improve $J_c(H)$, as has been found for the A15 compounds,[43] which are now widely used in superconducting magnets.

Studies of the superconducting properties as a function of x in $La_{2-x}Sr_xCu_{4-d}$ and as a function of oxygen content in $(RE)Ba_2Cu_3O_{7-\delta}$ yield information which places constraints on viable theories. An observation inferred from such studies in $La_{2-x}Sr_xCuO_{4-\delta}$ by Shafer, Penney, and Olson,[44] and subsequently confirmed for $YBa_{2-x}La_xCu_3O_{7-\delta}$,[33] is that the transition temperature T_c scales approximately linearly with the concentration of $[Cu-O]^+$ complexes. A linear relationship of the form

$$kT_c = \left(\frac{2\pi\hbar^2}{m^*}\right)(n - n_c) \qquad (2)$$

where n is the concentration of holes per unit area of a single Cu-O plane, is predicted by theories proposing a Bose-Einstein type condensation, such as the resonant valence bond (RVB) theory of Anderson et al.[25,26] Thus, if the $[Cu-O]^+$ complexes are identified with mobile holes, then Eq. (1) is qualitatively consistent with the observations of Shafer et al.[44] However, when the estimated effective mass of $m^* \sim 9m$[40] is substituted into Eq. (1), the predicted values of T_c are found to be about an order of magnitude higher than the values experimentally observed for a given hole concentration. A linear term observed in the temperature-dependent specific heat at low temperatures[45-53] is consistent with gapless theories, such as the RVB model,[25,26] but a linear term has also been observed in oxygen-depleted, nonsuperconducting samples of $YBa_2Cu_3O_{7-\delta}$,[54] suggesting that it may be due to glassy states and not directly related to the excitation spectrum of the superconducting condensate.

A further problem with gapless theories is the ample

experimental evidence supporting the existence of a BCS energy gap 2Δ, which indicates a BCS type pairing. Numerous FIR reflectivity[55-58] and tunneling[59-61] studies demonstrate the existence of an energy gap in the elementary excitation spectrum which has a BCS form as a function of temperature, although there is disagreement about the magnitude of the gap. Published values of $2\Delta(0)/k_B T_c$ range from ~2 to ~14. The most recent measurements of $2\Delta(0)/k_B T_c$ ~5-8 indicate that the high T_c superconductors are strongly coupled. Measurements[62] of the nuclear spin relaxation rate as a function of temperature demonstrate the existence of two BCS energy gaps in YBCO, one of which may be related to the Cu-O chains, and the other of which may correspond to the Cu-O planes.

According to theories which postulate pairing by the exchange of virtual electronic resonances (excitons, charge-transfer excitations, etc.), there may be optically active excitations which correlate with the superconducting properties. In order to test this hypothesis, Kamarás et al[63] have performed a Kramers-Kronig (K-K) analysis of reflectivity spectra of $La_{2-x}Sr_xCuO_4$ (LSCO) and $YBa_2Cu_3O_{7-\delta}$ (YBCO) ceramics. Their measurements on YBCO included a direct comparison of the reflectivity of a superconducting sample with that of a nonsuperconducting, oxygen-depleted, sample. The results of their K-K analysis indicates the presence of a broad peak in the optical conductivity centered about an energy of ~0.37 eV and a smaller peak at ~ 2 eV which are present only in the superconducting sample. Unfortunately, reflectivity measurements of ceramics, where the crystallites are randomly oriented, can sometimes lead to an ambiguous interpretation, and their straightforward K-K analysis has been criticized. One inconsistency in the analysis is that the limiting low frequency conductivity obtained from the K-K calculation is about a factor of five off from the measured dc value.

A recent K-K analysis of the reflectivity spectrum of a superconducting YBCO single crystal,[64] shows no peak in the optical conductivity spectrum in the range 0.1-1.0 eV. Similarly optical transmission measurements on oriented, polycrystalline YBCO thin films,[65] show no evidence for well-defined peaks in the optical absorption spectrum in the same energy range. The IBM[64] and Stanford[65] results do not completely resolve the issue of whether or not a resonance exists, however, for the following reason. According to the theories of Little,[19] Ginzburg,[20] and Allender *et al*,[21] the pairing of superconducting carriers in one layer (chain or plane) is mediated by the virtual polarization of electrons in an adjacent side chain or plane. This polarization will essentially be perpendicular to the superconducting (Cu-O) layer. Thus, in order to optically excite this resonance in YBCO, the electric field would have to be polarized along the c-axis. In the reflectivity studies conducted at IBM[64] and the optical transmission studies conducted at Stanford[65] the electric field was oriented along the **a-b** plane, where the metallic Drude response may have swamped any possible, probably small, resonance. Furthermore, it is possible that the virtual excitations which mediate the pairing have nonzero wavevector and are thus not optically active.[66]

Assuming, for the moment, that there are electronic or magnetic virtual excitations which mediate tha pairing process, it is not immediately clear how the linear dependence of T_c on the concentration of $[Cu-O]^+$ complexes would arise within the BCS framework. However, Newns[67] has recently been able to account for the observed scaling by postulating a two-particle resonance pairing mechanism, in which the pairing ineteraction results from virtual transitions between different copper d valence states. If x_h represents the average number of holes per copper atom, then, according to his weak-coupling calculations, T_c is predicted to scale as

$$T_c \propto \frac{x_h\left(1 - x_h^2\right)^{\frac{1}{2}}}{\varepsilon_d + x_h D} \tag{3}$$

where ε_d (~3.4 eV for La_2CuO_4) represents an effective d-band energy and 2D (~5 eV for La_2CuO_4) is about equal to the width of the oxygen p-band. According to equation (3), T_c should scale approximately linearly with x_h for $x_h < 0.3$, but should saturate and begin to decrease for x_h > 0.6. An important test of this theory would be to develop methods of further enhancing the hole concentration in doped La_2CuO_4 and "123"-type superconductors.

In summary, there are at present a great many more unknowns, including the nature of the pairing mechanism, than knowns for the high T_c superconducting perovskites. However, rapid progress is being made towards a detailed understanding of the microscopic origin of high T_c superconductivity. Eventually, an improved understanding is likely to lead to a variety of novel, yet undreamed of, applications.

ACKNOWLEDGMENTS

I wish to thank W. E. Hatfield, C. W. Chu, M. K. Wu, W. K. Chu, J. Bardeen, and others for stimulating conversations. I acknowledge receipt of an A. P. Sloan Research Fellowship, an R. J. Reynolds Junior Faculty Development Award, and a grant from the University of North Carolina at Chapel Hill.

REFERENCES

1. J. G. Bednorz and K. A. Müller, Z. Physik B**64**, 189 (1986).

2. S. Uchida, H. Tagaki, K. Kiazawa, and S. Tanaka, Jap. J. Appl. Phys. Lett. **26**, L1 (1987).

3. C. W. Chu, P. H. Hor, R. L. Mong, L. Gao, Z. J. Huang, and Y. Q. Wang, Phys. Rev. Lett. **58**, 405 (1987).

4. R. J. Cava, R. B. Van Dover, B. Batlogg, and E. A. Rietmann, Phys. Rev. Lett. **58**, 408 (1987).

5. M. K. Wu, J. R. Ashburn, C. J. Torng, P. H. Hor, R. L. Meng, L. Gao, Z. J. Huang, Y. Q. Wang, and C. W. Chu, *Phys. Rev. Lett.* **58**, 908 (1987).

6. P. H. Hor, L. Gao, R. L. Meng, Z. J. Huang, Y. Q. Wang, F. Forster, J. Vassilious, C. W. Chu, M. K. Wu, J. R. Ashburn, and C. J. Torng, *Phys. Rev. Lett.* **58**, 911 (1987).

7. R. J. Cava, B. Batlogg, R. B. Van Dover, D. W. Murphy, S. Sunshine, T. Siegrist, J. P. Remeika, E. A. Rietmann, S. Zahurak, and G. P. Espinosa, *Phys. Rev. Lett.*, **58**, 1676 (1987).

8. T. Yamashita *et al*, *Jap. J. of Appl. Phys.* **26**, L635 and L671 (1987).

9. R. H. Koch, C. P. Umbach, G. J. Clark, P. Chaudari, and R. B. Laibowitz, *Appl. Phys. Lett.* **51**, 200 (1987).

10. M. R. Schafroth, J. M. Blatt, and S. T. Butler, *Helv. Phys. Acta* **30**, 93 (1957); M. R. Schafroth, *Phys. Rev.* **111**, 72 (1958).

11. See, for example, J. R. Schrieffer, *Theory of Superconductivity*, Benjamin/Cummings, Reading, Mass. (1964,1983).

12. C. M. Varma and S. Schmitt-Rink, *Proc. of the International Workshop on Novel Mechanisms of Superconductivity*, S. A. Wolf and V. Z. Kresin, ed. Berkeley, California, June 22-24, 1987.

13. W. Weber, *Phys. Rev. Lett.* **58**, 1371 (1987).

14. K. Miyake, S. Schmitt-Rink, and C. M. Varma, *Phys. Rev.* **B34**, 6554 (1986).

15. D. J. Scalapino, E. Loh, and J. Hirsch, *Phys. Rev.* **B34**, 8190 (1986).

16. C. J. Pethick and David Pines, *Proc. of the International Workshop on Novel Mechanisms of Superconductivity*, S. A. Wolf and V. Z. Kresin, ed. Berkeley, California, June 22-24, 1987.

17. V. J. Emery, *Phys. Rev. Lett.* **58**, 2794 (1987).

18. R. H. Parmenter, *Phys. Rev. Lett.*, **59**, 923 (1987).

19. W. A. Little, *Phys. Rev.* **134**, A1416 (1964).

20. V. L Ginzburg, *Zh. Eksp. Teor. Fiz.* **47**, 2318 (1964) [*Sov. Phys. - JETP* **20**, 1549 (1965)].

21. D. Allender, J. Bray, and J. Bardeen, *Phys. Rev.* **B7**, 1020 (1973).

22. C. M. Varma, S. Schmitt-Rink, and E. Abrahams, *Solid State Commun.* **62**, 681 (1987).

23. V. Z. Kresin, *Phys. Rev.* **B35** (1987).

24. P. Prelovsek, T. M. Rice, and F. C. Zhang, *J. Phys.* **C 20**, L229 (1987).

25. P. W. Anderson, *Science* **235**, 1196 (1987).

26. P. W. Anderson, G. Baskaran, Z. Zhou, and T. Hsu, *Phys. Rev. Lett.* **58**, 2790 (1987).

27. B. Batlogg, A. P. Ramirez, R. J. Cava, R. B. van Dover, and E. A. Rietman, *Phys. Rev.* **B35**, 5340 (1987).

28. B. Batlogg, R. J. Cava, A. Jayaraman, R. B. van Dover, G. A. Kaurouklis, S. Sunshine, P. W. Murphy, L. W. Rupp, H. S. Chen, A. White, K. T. Short, A. M. Mujsce, and E. A. Rietman, *Phys. Rev.*

Lett. **58**, 2333 (1987).

29. L. C. Bourne, M. F. Crommie, A. Zettl, Hans-Conrad zur Loye, S. W. Keller, K. L. Leary, Angelica M. Stacy, K. J. Chang, Marvin L. Cohen, and Donald E. Morris, *Phys. Rev. Lett.* **58**, 2337 (1987).

30. B. Batlogg, G. Kourouklis, W. Weber, R. J. Cava, A. Jayaraman, A. E. White, K. T. Short, L. W. Rupp, and E. A. Rietman, *Phys. Rev. Lett.* **59**, 912 (1987).

31. Tanya A. Faltens, William K. Ham, Steven W. Keller, Kevin J. Leary, James N. Michels, Angelica M. Stacy, Hans-conrad zur Loye, Donald E. Morris, T. W. Barbee III, L. C. Bourne, Marvin L. Cohen, S. Hoen, and A. Zettl, *Phys. Rev. Lett.* **59**, 915 (1987).

32. L. C. Bourne, A. Zettl, T. W. Barbee III, and Marvin L. Cohen, *Phys. Rev.* **B36**, 3991 (1987).

33. B. Batlogg, *International Symposium on High Temperature Superconducting Materials*, University of North Carolina at Chapel Hill, Sept. 18-19, 1987.

34. M. Gurvitch and A. T. Fiory, *Phys. Rev. Lett.* **59**, 1337 (1987).

35. W. L. McMillan, *Phys. Rev.* **167**, 331 (1968);   W. L. McMillan and J. M. Rowell, *Phys. Rev. Lett.* **14**, 108 (1965).

36. J. Bardeen, L. N. Cooper, and J. R. Schrieffer, *Phys. Rev.* **106**, 162 (1957); **108**, 1175 (1957).

37. G. M. Eliashberg, *Zh. Eksperim. i Teor. Fiz.* **38**, 966 (1960) [*Soviet Phys. - JETP* **11**, 696 (1960)].

38. Y. C. Jean, S. J. Wang, H. Nakanishi, W. N. Hardy, M. E. Hayden, R. F. Kiefl, R. L. Meng, H. P. Hor, J. Z. Huang, and C. W. Chu,

*Phys. Rev. B***36**, 3994 (1987).

39. W. Y. Ching, Yongnian Xu, Guang-Lin Zhao, K. W. Wong, and F. Zandiehnadem, *Phys. Rev. Lett.* **59**, 1333 (1987).

40. J. Bardeen, D. M. Ginsberg, and M. B. Salamon, *Proc. of the International Workshop on Novel Mechanisms of Superconductivity*, S. A. Wolf and V. Z. Kresin, ed. Berkeley, California, June 22-24, 1987.

41. P. M. Grant, R. B. Beyers, E. M. Engler, G. Lim, S. S. P. Parkin, M. L. Ramirez, V. Y. Lee, A. Nazzal, J. E. Vazquez, and R. J. Savoy, *Phys. Rev. B*35, 7242 (1987).

42. Gang Xiao, F. H. Streitz, A. Garvin, M. Z. Cieplak, J. Childress, Ming Lu, A. Zwicker, and C. L. Chien, *Phys. Rev. B***36**, 2382 (1987).

43. M Suenaga, W. B. Sampson, and C. J. Klamut, *IEEE Trans. Mag.* **MAG-11**, 231 (1975).

44. M. W. Shafer, T. Penney, and B. L. Olson, *Phys. Rev. B***36**, 4047 (1987).

45. M. E. Reeves, T. A. Friedmann, and D. M. Ginsberg, *Phys. Rev. B***35**, 7207 (1987).

46. A. P Ramirez, B. Batlogg, G. Aeppli, R. J. Cava, E. Rietman, A. Goldman, and G. Shirane, Phys. Rev. B35, 8833 (1987).

47. L. E. Wenger, J. T. Chen, G. W. Hunter, and E. M. Logothetis, *Phys. Rev. B***35**, 7213 (1987).

48. Z. Zirngiebl, J. O. Willis, J. D. Thompson, C. Y. Huang, J. L. Smith, P. H. Hor, R. L. Meng, C. W. Chu, and M. K. Wu, preprint.

49. A. Junod, A Bezinge, D. Cattani, J. Cors, M. Decroux, O. Fischer, P. Genoud, L. Hoffmann, J. L. Jorda, J. Muller, and E. Walker, preprint.

50. S. E. Inderhees, M. B. Salamon, T. A. Friedmann, and D. M. Ginsberg, *Phys. Rev. B*36, 2401 (1987).

51. A. Junod, A. Bezinge, T. Graf, J. L. Jorda, J. Muller, L. Antognazzi, D. Cattani, J. Cors, M. Decroux, O. Fischer, M. Banovski, P. Genoud, L. Hoffmann, A. A. Manuel, M. Peter, E. Walker, M. Francois, and K. Yvon, preprint.

52. B. D. Dunlap, M. V. Nevitt, M. Slaski, T. E. Klippert, Z. Sungaila, A. G. McKale, D. W. Capone, R. B. Poeppel, and B. K. Flandermeyer, *Phys. Rev. B*35, 7210 (1987).

53. M. E. Reeves, D. Citrin, B. G. Pazol, T. A. Friedmann, and D. M. Ginsberg, *Phys. Rev. B*36, 6915 (1987).

54. D. Haase and A. Kingon, preprint.

55. U. Walter, M. S. Sherwin, A. Stacy, P. L. Richards, and A. Zettl, *Phys. Rev. B*35, 5327 (1987).

56. P. E. Sulewski, A. J. Sievers, S. E. Russek, H. D. Hallen, D. K. Lathrop, and R. A. Buhrman, *Phys. Rev. B*35, 5330 (1987).

57. Z. Schlesinger, R. L. Greene, J. G. Bednorz, and K. A. Müller, *Phys. Rev. B*35, 5334 (1987).

58. Z. Schlesinger, R. T. Collins, and M. W. Shafer, *Phys. Rev. B*1 35, 7232 (1987).

59. I. Iguchi, H. Watanabe, Y. Kasai, T. Mochiku, A. Sugishita, and

E. Yamaka, *Jap. J. Appl. Phys.* **26**, L645 (1987).

60. J. R. Kirtley, C. C. Tsuei, Sung I. Park, C. C. Chi, J. Rozen, and M. W. Shafer, *Phys. Rev. B1* **35**, 7216 (1987).

61. J. R. Kirtley, R. T. Collins, Z. Schlesinger, W. J. Gallagher, R. L. Sandstrom, T. R. Dinger, and D. A. Chance, *Phys. Rev. B35*, 8846 (1987).

62. W. W. Warren, Jr., R. E. Walstedt, G. F. Brennert, G. P. Espinosa, and J. P. Remeika, *Phys. Rev. Lett.* **59**, 1860 (1987).

63. K. Kamarás, C. D. Porter, M. G. Doss, S. L. Herr, D. B. Tanner, D. A. Bonn, J. E. Greedan, A. H. O'Reilly, C. V. Stager, and T. Timusk, *Phys. Rev. Lett.* **59**, 919 (1987).

64. Z. Schlesinger, R. T. Collins, D. L. Kaiser, and F. Holtzberg, *Phys. Rev. Lett.* **59**, 1958 (1987).

65. I. Bozovic, D. Kirillov, A. Kapitulnik, K. Char, M. R. Hahn, M. R. Beasley, T. H. Geballe, Y. H. Kim, and A. J. Heeger, *Phys. Rev. Lett.* **59**, 2219 (1987).

66. J. Bardeen, personal communication.

67. D. M. Newns, *Phys. Rev. B36*, 5595 (1987).

5

Density Functional Theory for Hidden High-T$_c$ Superconductivity

Akitomo Tachibana   Department of Chemistry, University of North Carolina, Chapel Hill, North Carolina 27514

1. INTRODUCTION

Recent revolutionary findings of novel oxide superconductors [1] promote a renewed detailed study of the theory of superconductivity [2]. Bardeen, Cooper and Schrieffer (BCS) have established a general statistical mechanical theory for superconducting phase transition, emphasizing electron-phonon coupling as the underlying mechanism for the attractive Cooper pairing force [3]. Others also follow this line, emphasizing electronic and magnetic couplings as the underlying mechanism, and construct the electronic correlation models responsible for superconductivity [2,4-13]. Also, recently the electron-phonon coupling has been endowed with a new feature in which the vibronic attractive force for Anderson time-reversal pairing occurs in the unperturbed Hamiltonian [14].

    In this paper, we generalize the statistical mechanical framework of the BCS theory, with the result that the BCS vacuum is polarized [15]. A new superconducting phase appears at Tb, higher than the conventional well-defined critical temperature Tc. This appears to explain the recent finding of novel superconductors at higher temperature. The density functional theory [16,17] is used, with no recourse to any model Hamiltonian.

2. QUASI-PARTICLE IN THE FERMI LIQUID

Density functional theory gives a systematic unified picture of the electronic structure of materials, from atoms and molecules to condensed matters in general [16-22].

We are dealing with large but finite electrically neutral materials for which superconductivity is observed. The electronic Hamiltonian includes effects of the surface boundary, as we shall show below.

Inside the material, the attractive Coulombic potentials of nuclei are smoothed by superposition, with the result that a gigantic Fermi sea of electrons is formed. The nearly free electron model for valence and conducting electrons is applicable with great accuracy. Along the saeshore, however, a high cliff of potential barrier is left, with the result that it prevents the electrons from flowing outside the sea. The work function barrier manifests the effects of the surface boundary and gives an electron the asymptotic potential-energy value which is set equal to zero. Thus the energy of each electron, or quasi-particle in general, should be negative with respect to the horizon, insofar as the system is stable against the removal of electrons from the material, leaving a positive ionic core system behind. The stability is also guaranteed by Le Chatlier's principle for the thermal equilibrium state.

The best possible choice of the exact wave function $\phi_{i\sigma}(\vec{r})$ for the quasi-particle is provided by the density functional theory, as the eigenfunction of the common operator $L_\sigma(\vec{r})$ [15],

$$L_\sigma(\vec{r})\phi_{i\sigma}(\vec{r}) = e_{i\sigma}\phi_{i\sigma}(\vec{r}) \quad ; \quad <\phi_{i\sigma}|\phi_{j\sigma}> = \delta_{ij} \quad , \tag{1}$$

$$L_\sigma(\vec{r}) = -(1/2)\Delta(\vec{r}) + \int[\partial F/\partial\nu(\vec{r}')]\chi_{\nu\rho_\sigma}(\vec{r}',\vec{r})d\vec{r}' \quad , \tag{2}$$

$$\chi_{\nu\rho_\sigma}(\vec{r}',\vec{r}) = \delta\nu(\vec{r}')/\delta\rho_\sigma(\vec{r}) \quad . \tag{3}$$

Here F is the universal functional of electron charge density $\rho(\vec{r})$, $\nu(\vec{r})$ is the Lagrange multiplier adopted in the "apparatus" density functional theory [22] for the variational constraint of fixed $\rho(\vec{r})$, and where $\chi_{\nu\rho_\sigma}(\vec{r}',\vec{r})$ is the response function of $\nu(\vec{r}')$ with respect

to the spin σ-component $\rho_\sigma(\vec{r})$ of $\rho(\vec{r})$. The serial number i is unique
for each spin. For example, $(\phi_{i\alpha}(\vec{r}),\phi_{i\beta}(\vec{r}))$ denotes the pair with
α-spin plane wave with momentum $\vec{k}$, and β-spin plane wave with
momentum $-\vec{k}$, a Cooper pair. The eigenvalue $e_{i\sigma}$ is also given as

$$e_{i\sigma} = \partial E/\partial f_{i\sigma} \quad , \tag{4}$$

where E is the internal energy of the whole system, electrons and
nuclei, and where $f_{i\sigma}$ is the occupation number of the quasi-particle.

We shall treat the superconducting phase transition of the
Fermi liquid, dictating the Bogoliubov transformation

$$\hat{a}_{i\sigma}^+ = (\hat{c}_{i\sigma}^+ + g_i^*\hat{c}_{i-\sigma})/(1 + |g_i|^2)^{1/2} \quad ; \quad g_{i\alpha}=-g_{i\beta}\equiv g_i \quad . \tag{5}$$

Here $\hat{a}_{i\sigma}^+$ and $\hat{c}_{i\sigma}^+$ denote the creation operators of the quasi-particle
and the superconducting pairing state, respectively. The occupation
number of the latter is given as

$$d_{i\sigma} = <\hat{c}_{i\sigma}^+\hat{c}_{i\sigma}> \quad , \tag{6}$$

where $<\,>$ denotes the trace with respect to the density matrix. We
find the relationship, using a real weight function w [15],

$$<\hat{c}_{i\beta}^+\hat{c}_{i\alpha}^+> = (1/2)w<\hat{a}_{i\beta}^+\hat{a}_{i\alpha}^+> \quad , \tag{7}$$

which manifests the BCS vacuum polarization. In the original BCS
theory the quantity $<\hat{c}_{i\beta}^+\hat{c}_{i\alpha}^+>$ is zero. The consequence of it not being
zero in general is that the upper limit of temperature for which
nonzero g_i is found is Tb, higher than the conventional critical
temperature Tc at which a positive jump (with decreasing T) of the
electronic specific heat is found, as we shall show in the following
section. The weight function itself is obtained as [15]

$$w = 2(k-x)/(k+x) \quad , \tag{8}$$

where k is a real, explicit function of $d_{i\sigma}$, and hence is an implicit
function of x through $d_{i\sigma}$. The k has the meaning of the ratio of the
pairing state mixing,

$$f_{i\sigma} = \{d_{i\sigma}+k(1-d_{i-\sigma})\}/(1+k) \quad . \tag{9}$$

The distribution of the zeros of w is very interesting, because the
BCS theory requires w = 0 identically. Using Eq.(8), the equation
$k(d_{i\sigma}(x)) = x$ determines the zeros of w. The deviation of k from x

should be small, because the polarization of the BCS vacuum is supposed to be small.

3.   SPECIFIC HEAT AND SPIN SUSCEPTIBILITY

The $d_{i\sigma}$ is given as [15]

$$d_{i\sigma} = 1/\{e^{\beta(E_{i\sigma} - \frac{1-x}{1+x}\mu_G)} + 1\} \quad , \tag{10}$$

$$E_{i\sigma} = \partial E/\partial d_{i\sigma}$$

$$= \frac{1}{1+x}e_{i\sigma} - \frac{x}{1+x}e_{i-\sigma} \quad ; \quad x = |g_i|^2 \quad , \tag{11}$$

where we put $w = 0$, an assumption loosened in the next section. μ_G is the Gibbs chemical potential. It should be noted that the chemical potential for the pairing state is modified, and hence the energy spectrum which contributes to the electronic specific heat has the extra x-dependence over and above that of $E_{i\sigma}$. Since μ_G is negative definite for finite stable equilibrium system, the above feature is significant, as we shall show below.

For particles for which $e_{i\sigma} - \mu_G > 0$, a positive binding energy is found [15],

$$(e_{i\sigma}-\mu_G)-(E_{i\sigma} - \frac{1-x}{1+x}\mu_G) = \frac{x}{1+x}(e_{i\sigma}-\mu_G+e_{i-\sigma}-\mu_G) > 0 \quad . \tag{12}$$

This is a monotone increasing function of x. If x exceeds $x_{ci\sigma}$,

$$x_{ci\sigma} = (e_{i\sigma}-\mu_G)/(e_{i-\sigma}-\mu_G) \quad , \tag{13}$$

then the energy becomes negative,

$$E_{i\sigma} - \frac{1-x}{1+x}\mu_G < 0 \quad \text{if} \quad x > x_{ci\sigma} \quad . \tag{14}$$

The critical point defines Tc. For a Cooper pair, $e_{i\sigma}=e_{i-\sigma}$, we have

$$x_{ci\sigma} = 1 \quad \text{at} \quad T = Tc \quad \text{if} \quad e_{i\sigma} = e_{i-\sigma} \quad . \tag{15}$$

In general, the contribution Ces to the electronic specific heat Ce is given by [15]

$$Ces = \sum_{\sigma i}\sum (E_{i\sigma} - \frac{1-x}{1+x}\mu_G)d_{i\sigma}(1-d_{i\sigma})$$

$$\times \{(E_{i\sigma} - \frac{1-x}{1+x}\mu_G)/k_B T^2 - (d/dT)(E_{i\sigma} - \frac{1-x}{1+x}\mu_G)/k_B T\} \quad . \tag{16}$$

Assuming the phase transition to be second order, the Ce experiences negative jump at Tb, in contrast to the positive jump at Tc, the latter being the conventional behavior. This behavior holds true if holes, for which $e_{i\sigma} - \mu_G < 0$, contribute to superconducting phase transition. The above property should be observed by direct measurement of Ce [23]. If higher-order phase transition occurs at Tb, then the discontinuity in Ces will not be observed at Tb.

Furthermore, we can prove the difference rule [15],

$$f_{i\alpha} - f_{i\beta} = d_{i\alpha} - d_{i\beta} \ .$$ (17)

If the difference becomes zero at T=0, then the spin susceptibility χ becomes zero. Hence the Knight shift, if present, should be caused by spin-orbit interactions in the surface region [24,25]. There is a possibility, however, that the value in Eq.(17) is not zero at T=0, which is allowed in the present theory [15], and the nonzero χ may lead to a Knight shift through Fermi contact interaction. This is a variant of the Heine-Pippard mechanism [26].

4. ENERGY GAP

The energy spectrum for the quasi-particle in the superconducting phase is given by Eq.(10) as

$$E = E_{i\sigma} - \frac{1-x}{1+x} \mu_G \ .$$ (18)

This is valid when the polarization effect is negligible. At T = Tc the energy crosses the E = 0 line. An interaction may emerge at the crossing point. As a matter of fact, (a) the paired state at the Fermi level and (b) the state of configuration in which a particle is created (or annihilated) and the mirror-image hole is annihilated (or created), should have the same energy. Hence, there emerges a strong interaction associated with the pair-breaking (or recombination) process. This is observed in the Giaever tunneling experiment [23]. Then we require $w \neq 0$ and obtain the renormalized excitation energy for the quasi-particle as [15]

$$E = (\epsilon^2 + \Delta^2)^{1/2} \ ,$$ (19)

where we replace Eq.(18) by the same energy ϵ of the reference normal electron, and where Δ is the energy gap. Especially for a

Cooper pair, $e_{i\sigma} = e_{i-\sigma}$, we then obtain $x = \{1-\varepsilon/(e_{i\sigma}-\mu_G)\}/$
$\{1+\varepsilon/(e_{i\sigma}-\mu_G)\}$. Comparing this expression of x with the BCS
expression, we find the renormalization effect in $e_{i\sigma}$ as $|e_{i\sigma}-\mu_G| =$
$(\varepsilon^2+\Delta^2)^{1/2}$. Mathematically, the renormalization effect in $e_{i\sigma}$ near
T = Tc should be the consequence of the exact secular equation for
$\phi_{i\sigma}$, Eqs.(1)-(3). The $\phi_{i\sigma}$ for $e_{i\sigma}$ is then considered the wave
function into which the paired state is decomposed.

If the pairing state is formed only in the very narrow energy
shell centered at the Fermi level, i.e. $|e_{i\sigma} - \mu_G| \ll 1$, and if Tb
is close to Tc, then the avoided crossing smears out the fine
structure at T = Tb.   This is exactly the situation that may occur
for the conventional old type of superconductors. The consequence
is that the negative jump disappear, there remaining only the
positive jump at T = Tc.   When this is not the case, such a
phenomenon could be observed in a quite unusual circumstance, such
as in the recently found novel high-Tc superconductors. If we
further assume that Tb and Tc coalesce, then we recover the original
picture of the superconducting phase transition established by the
BCS theory.

In general, asymptotically as x goes to zero, the k in Eq.(8)
should go to zero and the difference should go to zero faster than
x itself does, because the polarization of the BCS vacuum should be
small. So, we recover Eq.(18).

5. DISCUSSION

The basic difference between the present theory and the original
BCS theory are two. First, the present theory uses no model
Hamiltonian, with the consequence that Eq.(7) holds and has nonzero
value whenever the g_i in Eq.(5) is nonzero. This manifests the BCS
vacuum polarization [15]. Second, the chemical potential μ_G is
nonzero in the present theory for the reasons stated at the
beginning of this paper, and the modification of it in the energy
spectrum for the quasi-particle in the superconducting phase plays
an important role for the anomaly in Ces, while it is used merely a
zero of energy in the BCS theory. We have thus generalized the
statistical mechanical framework of the original BCS theory.

The underlying mechanism of the attractive pairing force has not
yet met general consensus. In this connection, it should be noted
that strong electron-phonon interaction is brought about by the
Jahn-Teller effects in the novel oxide superconductors. This can go
beyond the original BCS theory for the attractive pairing force [27].
This may cause the fluctuation or polarization of the BCS vacuum.
Indeed, the exact vibronic Hamiltonian manifests the attractive
pairing force if the vibronic interaction is very large [14].
In addition to the strong vibronic interaction, other underlying
mechanisms may also play important roles [2,4-13].

ACKNOWLEDGEMENTS
The author wishes to thank Professor Robert G. Parr for his kind
discussion, encouragement and help with the manuscript. This work
is supported by grants from the National Science Foundation and the
National Institutes of Health to the University of North Carolina.

(a) Permanent address: Department of Hydrocarbon Chemistry, Faculty
of Engineering, Kyoto University, Kyoto 606, Japan; Division of
Molecular Engineering, Faculty of Engineering, Kyoto University,
Kyoto 606, Japan.

REFERENCES
1.   J.G. Bednorz and K.A. Müller, Z. Phys. B64, 188 (1986); S.
     Uchida, H. Takagi, K. Kitazawa and S. Tanaka, Jpn. J. Appl.
     Phys. Part 2 26, L1 (1987); C.W. Chu, P.H. Hor, R.L. Meng, L.
     Gao, Z.J. Huang and Y.Q. Wang, Phys. Rev. Lett. 58, 405 (1987).
2.   T.M. Rice, Z. Phys. B67, 141 (1987); and references cited
     therein.
3.   J. Bardeen, L.N. Cooper and J.R. Schrieffer, Phys. Rev. 106,
     162 (1957); 108, 1175 (1957).
4.   W.A. Little, Phys. Rev. 134, A1416 (1964).
5.   V.L. Ginzburg, Zh. Eksp. Teor. Fiz. 47, 2318 (1964). [Sov. Phys.
     JETP 20, 1549 (1965)].
6.   P.W. Anderson, Mater. Res. Bull. 8, 153 (1973); P.W. Anderson,
     G. Baskaran, Z. Zou and T. Hsu, Phys. Rev. Lett. 59, 225 (1987).

7. L. Pauling, Proc. Natl. Acad. Aci. USA 60, 59 (1968); Phys. Rev.
    Lett. 59, 225 (1987).

8. S. Kivelson, D. Roskhsar and J. Sethna, Phys. Rev. B35, 8865
    (1987).

9. L.F. Mattheiss, Phys. Rev. Lett. 58, 1028 (1987).

10. D.M. Newns, preprint.

11. V.J. Emery, Phys. Rev. Lett. 58, 2794 (1987).

12. R.H. Parmenter, Phys. Rev. Lett. 59, 923 (1987).

13. C.M. Varma, S. Schmitt-Rink and E. Abrahams, Solid State
    Commun. 62, 681 (1987).

14. A. Tachibana, Phys. Rev. A35, 18 (1987).

15. A. Tachibana, Phys. Rev. Lett., submitted (received 27 August
    1987); paper in preparation.

16. P. Hohenberg and W. Kohn, Phys. Rev. 136, B864 (1964).

17. N.D. Mermin, Phys. Rev. 137, A1441 (1965).

18. R.G. Parr, Ann. Rev. Phys. Chem. 34, 631 (1983); in Local
    Density Approximation in Quantum Chemistry and Solid State
    Physics, ed. by J.P. Dahl and J. Avery (Plenum, NY, 1984),
    p. 21; in Density Functional Methods in Physics, ed. by R.M.
    Dreizler and J. da Providencia (Plenum, NY, 1985), p. 141;
    J.F. Capitani, R.F. Nalewajski and R.G. Parr, J. Chem. Phys.
    76, 568 (1982).

19. W. Kohn and P. Vashishta, in Theory of the Inhomogeneous
    Electron Gas, ed. by S. Lundqvist and N.H. March (Plenum, NY,
    1983), p. 79.

20. U. von Barth, in Many-Body Phenomena at Surfaces, ed. by D.
    Langreth and H. Suhl (Academic, NY, 1984), p. 3.

21. J.P. Perdew, Int. J. Quant. Chem. Quant. Chem. Symp. 19, 497
    (1986).

22. A. Tachibana, Phys. Rev. A, submitted.

23. Superconductivity, ed. by R.D. Parks (Marcel Dekker, NY, 1969).

24. R.A. Ferrell, Phys. Rev. Lett. 3, 262 (1959).

25. P.W. Anderson, Phys. Rev. Lett. 3, 325 (1959).

26. V. Heine and A.B. Pippard, Phil. Mag. 3, 1046 (1958).

27. K.A. Müller and J.G. Bednorz, Science 237, 1133 (1987).

6

Where Can New Classes of High-T_c Superconducting Materials Be Found?

FRANCIS J. DISALVO   Cornell University, Ithaca, New York 14853

At present (Sept. 1987), there are several known oxide compounds that are superconducting at temperatures above 30K, with some approaching 100K, and numerous reports of metastable drops in resistance in some materials at temperatures as high as room temperature. There are two types of copper oxides in this class with different but related crystal structures: $La_{2-x}M_xCuO_4$ with M = Ca, Sr, or Ba [1,2]; $MBa_2Cu_3O_7$ with M = trivalent metals [3-6]. ($La_{3-x}Ba_{3+x}Cu_6O_{14}$ is closely related to $MBa_2Cu_3O_7$ [7].) The structures are derivatives of perovskite, which has the composition $MM'O_3$, and are variously described as having square planar coordinated copper-oxygen sheets or copper in highly distorted octahedral or square pentagonal oxygen coordination. In $MBa_2Cu_3O_7$ there are also one dimensional Cu-O chains with Cu in square planar coordination. Another oxide, $BaPb_{1-x}Bi_xO3$, should probably also be included in this class, even though its T_c is only 13K, since on the basis of its density of states, its T_c should only be a few degrees [8]. Perhaps coincidently, it also has the perovskite structure.

Since the end of 1986, synthetic work has focused on related oxides obtained by substitution of the cations or anions or by an Edisonian variation of the composition. These approaches are an important process in the search for new high T_c materials. If general characteristics of the high T_c phases can be used as a guide to future synthesis strategies, it is hoped that a more rapid route to the discovery of other high T_c phases can result. Such a search presupposes a "faith" that the known oxide superconductors are not unique but that other compounds with similar or enhanced properties must exist. In this paper I outline some ideas on the unusual features of the oxide superconductors and some thoughts on what other compounds might show similar characteristics.

I start with a "disclaimer": while my students and I have prepared some of the high T_c materials by following others recipes and are just now starting to attempt to prepare new materials, I have not been one of the contributors to the large body of information that exists concerning these oxides and from which my thoughts arise. Further, I have been fortunate to receive many of the preprints that active groups have been circulating to avoid the "slowness" of the normal publication procedures. Without the openness of others, I would have little to say. I have also somewhat arbitrarily chosen the particular references to cite, choosing to pick one that illustrates the point, rather than trying to give an exhaustive list of all the preprints that I have seen on the topic.

The first and most surprising feature of these materials is apparent in band structure calculations. Such calculations may in fact not be correct in detail, since these materials appear to be close to the composition at which a Mott transition to the insulating state takes place [9]. This suggests that Coulomb correlations must be included in a realistic description of the properties. However, the observation based on the band structure calculations [10-12] that the states near the Fermi level have a strongly mixed character (of Cu d and O p states) will also likely survive in a "correct" description of the electronic structure. This admixture arises from the closeness in energy of the corresponding _atomic_ energy levels and from the fact that the band at the Fermi level is derived from a sigma antibonding state. This is indeed rather unusual for a metallic compound. The band structure results suggest that 50% or more of the wavefunction amplitudes are based on the oxygen! Perhaps, then, it is better to describe these materials as "metallic oxygen". In the vast majority of other metallic compounds the wavefunctions at the Fermi level are predominantly of cation (metal) character, typically containing 20% or less anion character. In other conducting oxides, particularly of early transition metals (such as $LiTi_2O_4$), the states at the Fermi level are derived from non-bonding or weakly pi bonding levels, further reducing the already low oxygen character resulting from the large energy difference between the oxygen p and titanium d states.

This large anion character in the wavefunctions also occurs in a few other non-superconducting oxides, such as PbO_2. Indeed, two new copper oxides, $La_4BaCu_5O_{13}$ and $La_{1+x}Ba_{2-x}Cu_3O_7$ (0<x<0.5)[13], have structures closely related to $YBa_2Cu_3O_7$ and are expected to have large oxygen contributions to the wavefunctions at the Fermi level (the band calculations have not yet been done to my knowledge). Yet these phases are only metallic and not superconducting down to 4.2K (for the second copper oxide listed above this is true for x>0.4). It is apparent then that a large

oxygen content in the wavefunctions at the Fermi level
is not by itself a <u>sufficient</u> condition for high T_c.
However, I believe that it is likely to found to be a
necessary condition.

There are several ways to describe the electronic
state of the materials. Band structure is one way and
formal valence states is another. While the latter is
oversimplified, it is also very easy and widely used.
But it can be a little misleading. For example, the
formal valence description of $YBa_2Cu_3O_7$ is
$Y^{+3}Ba^{+2}{}_2Cu^{+2}{}_2Cu^{+3}O^{-2}{}_7$. This implies that the oxygen p
levels are fully occupied (i.e., that the oxygen is
fully divalent). Since the oxygen and copper levels
each make a considerable contribution to the
wavefunctions at the Fermi level, it is impossible to
oxidize the copper to +3 without also oxidizing the
oxygen. In fact some recent experiments can detect only
Cu^{+2}, suggesting that only the oxygen is oxidized. In
that case the fraction of carrier density that is based
on oxygen is even higher than the band structure
calculations would suggest. Therefore, when authors
speak of needing "mixed valence" Cu^{+2}/Cu^{+3} in the
oxides to produce superconductivity, a broader
interpretation of what species are being oxidized
should be kept in mind.

The calculations also suggest that the
electropositive cations (alkaline earth or rare earth
metals) have little influence on the electronic
structure near the Fermi level. I believe that their
role is two fold. First, they help "enforce" a
particular structure; that is, the large cations are
responsible for the compounds adopting the perovskite
structure. Second, the electropositive metals
effectively increase the oxidizing power of the oxygen.
In binary copper oxides, the maximum formal valence

that can be obtained using the usual high temperature
solid state preparation techniques is +2. Whereas in
ternary copper oxides containing, for example, alkali
metals, copper can be formally +3, as in $NaCuO_2$.

The next unusual feature is that they are lousy
normal state metals. Their resistivity at 300K is
between 10^{-3} and 10^{-4} Ohm-cm, two to three orders of
magnitude higher than "good" metals and at least a few
factors higher than intermetallic transition metal
compounds [4,9,14]. This implies rather short mean
free paths for the conduction electrons, perhaps as
short as a lattice parameter. Yet the resistivity is
quite temperature dependent, typically varying linearly
with temperature. It is likely that the unusual
behavior of the resistivity is related to the nearness
to the Mott transition and may be an important key in
determining the detailed mechanism that leads to the
high T$_c$. At the same time the mechanical properties are
more like those of ceramics (brittle) than metals
(ductile). Recall, however, that the intermetallic
compounds are also often brittle, and that some
techniques have been invented to allow flexible wires
to be fabricated from them.

These materials are also fast ion conductors; at
least oxygen is able to diffuse rather readily through
the bulk of the compound at temperatures as low as
300C. This implies a bonding potential vs position for
oxygen in the lattice that is rather flat, or at least
barriers that are rather small in the direction of
diffusion. This in turn implies that oxygen vibrations
at lower temperatures will have a large amplitude,
perhaps being rather anharmonic. Recent neutron
diffraction measurements indeed show that the
crystallographically unique oxygen atoms in the "one
dimensional" chains in $YBa_2Cu_3O_7$ have large thermal

amplitudes of motion [5] and are the sites that are
reduced in occupancy when oxygen is removed from the
lattice [15]. (Some others have also suggested that the
fact that Cu^{+2} is a Jahn-Teller ion will also make its
motion rather anharmonic. However, the copper is
already in a rather distorted site, and the neutron
diffraction measurements show only a normal vibrational
amplitude.) This ready diffusion also has important
consequences for the synthesis and processing of the
material.

Some attention has been drawn to the low-
dimensional aspects of these materials. While the high
T_c materials have two dimensional sheets of Cu-O, and
the 90K superconductor has in addition one dimensional
Cu-O chains, $Ba(Pb/Bi)O_3$ has an almost cubic structure.
Consequently, low dimensionality seems not to be a
necessary condition for materials in this class to be
superconducting. In any case low dimensional structures
will likely result in other materials from low
coordination number of the transition metal and by
including large electropositive cations.

Before discussing some rather straight forward
generalizations of the above properties that might be
used as a guide for new synthesis, I want to say a few
words about the theories that have been proposed to
explain the high T_c's observed. While no theory has
been developed to the point that it explains in
microscopic detail all the observed phenomena, it might
be useful to take a brief look at them to extract
essential features. Then using these features and the
above observations, we can try to predict what new
compounds might be synthesized that reproduce some or
even all of the desired aspects of the known high T_c
materials.

I will limit this discussion, both since I am not

a theorist and because other such reviews exist [16].
The theories can be divided into three types, each
characterized by the principle interaction responsible
for the high T$_c$: phonon, magnetic, or exciton. The
phonon mechanisms might be divided into two categories:
enhanced electron- phonon interaction of the BCS type
[10] and local distortions about carriers to produce
bipolarons [17]. Both rely on the large coupling of the
electron energy to lattice motions that result from the
large mixing of the cation and anion orbitals and the
antibonding character of the wavefunctions at the Fermi
level. The magnetic models may also be grouped into two
categories: interaction with diffuse magnon modes and
interaction on a more local level to produce localized
singlet pairs (the RVB state, Resonating Valence Bond)
[18,19]. These theories emphasize the nearness of the
Mott insulating state to produce localized spin 1/2
Cu^{2+} sites. Some theories emphasize the large oxygen
contribution to the conductivity and the possibly local
nature of the electrons on copper [20]. Finally, the
exciton theories are based on a direct coupling of the
electrons to an electronic excitation. These might
include plasmons [21] or charge transfer excitations
[22]. The latter mechanism also relies on the small
energy difference between the cation d and anion p
states and on a large oscillator strength of the
transition. Each theory then picks one aspect of the
many unusual properties and attempts to explain the
high T$_c$ based on that feature.

    Now we can try to put this all together to try to
predict where else to look for new superconducting
phases. The main themes that arise from the preceding
discussion can be condensed into four general
characteristics: 1) A large cation-anion mixing of the
wavefunctions near the Fermi level, 2) metallic

conductor, but close to a Mott transition, 3)fast anion
conductor, 4) the electropositive cations do not play
an essential role in the electronic properties. For the
sake of discussion, we can arbitrarily break these
materials into oxides and other anion compounds.

Oxides

The atomic d states of copper are about 1eV above the p
states of oxygen. On moving to the left in the periodic
table, the d states rise by about 3eV when at Ti. The
energy of these levels are shifted when the atom is
incorporated in a compound. The anion levels rise and
the cation levels fall in energy, due to the charge
transfer and configuration mixing inherent in compound
formation. The charge transfer and energy shifts depend
upon the electronegativity difference between the anion
and cation and upon the cation-anion ratio. However,
the real charge transfer is usually considerably
smaller than that suggested by the formal valence.
Unfortunately, these shifts are difficult to predict
before a compound is even prepared. We can rely on
general trends to "guess" what will happen if the
cation is different than copper. If compounds are
prepared from 3d elements to the left of copper the
cation-mixing will tend to decrease, unless the
transition metal is in a higher average oxidation state
than that obtained in the copper compounds (which is
about +2.3). Perhaps if nickel compounds with an
average valence of greater than +3 could be prepared,
the mixing would again be strong. Using the same
reasoning it is possible that other late transition
elements will also be suitable. These include Ag, Pd,
Co, and Pt. For each element it may require special
preparation conditions to obtain the "correct" phases.

That is, unusual conditions such as preparation under high pressure or at low temperatures using solution techniques may be necessary.

In oxides, condition 2 (near a Mott transition) is likely to be met, if the compound is conducting at all. Most oxides are in fact not metallic conductors, and, of those that are, many exhibit metal insulator transitions with changing temperature or pressure. It may also be necessary that in the insulating state the cation has a spin 1/2 configuration (probably necessary for the RVB mechanism). The occurrence of the state depends upon the valence of the cation and upon its local coordination, with highly distorted near-neighbor environments favoring non degenerate states and spin 1/2 configuration . Alternatively, for a given environment the cation valence necessary to produce a spin 1/2 ion can easily be determined in the low and high crystal field strength limits.

It is not clear to me how to design a material to be a fast ion (anion) conductor, nor is it clear that this is a necessary condition for high T$_c$ superconductivity. If the large oxygen vibrations that are a consequence of the flat elastic potential for vibration produce an enhanced electron-phonon interaction, it would seem that such large vibrations could occur without ionic diffusion. Perhaps the flat elastic potential is a result of the near degeneracy in the cation and anion d and p levels. This would make charge transfer between them a low energy process, resulting in a high lattice polarizability, or equivalently, easily deformable ions. Such a picture is usually invoked in discussing fast ion conductors. So it could be that most of the systems that are potentially interesting superconductors of this type will also coincidentally be superionic conductors.

While a continued search for copper oxides containing mixed Cu^{+2}/Cu^{+3} formal valence states will be extended, perhaps a single example of another compound that should reproduce all the above features will suffice for illustrative purposes. Consider an oxide containing Ni^{+3}/Ni^{+4}. The formal valence state is higher than copper, so the Ni d states should be pulled down in energy, hopefully enough to again produce strong mixing with the oxygen. The electron configuration is d^7, which in a distorted cubic or square planar environment will be spin 1/2. A tetrahedral Ni environment in the low crystal field limit produces a spin 3/2 state, but in the large crystal field limit a spin 1/2 state would again be produced. If the compound is metallic, the band states at the Fermi level would be anti-bonding, thus further increasing the mixing and producing large electron-phonon coupling. Such Ni oxides would therefore be almost exact analogues of the high T_c copper oxides. Experimentally, the preparation of nickel oxides with an average formal valence greater than +3 may require high oxygen pressures or low temperature techniques.

Other anions

Other anions that are likely candidates include nitrogen, sulfur, and chlorine. Fluorine is unlikely to produce considerable mixing, because the p levels are a few volts below those of oxygen. Further, fluorides tend to be rather ionic and are very likely to be insulators (without a literature search, I can only think of a few conducting fluorides, Ag_2F and graphite intercalated with fluorine or fluorine containing anions). Mixing anions may produce small shifts in the cation and other anion levels and may be a suitable way to fine tune the properties. But the anion p levels are sufficiently different in energy that it is unlikely that both anions would mix with the cation levels to the same extent; one would always dominate.

Nitrides and sulfides that meet the conditions necessary for extensive mixing of the states at the Fermi level may be prepared if the cations are chosen from elements to the left of copper, since the anion p states are higher in energy than those of oxygen by about 1.5eV. If it is important to be near a metal-insulator transition, then choosing a compound based on a 3d transition element will enhance that probability. Empirically, such transitions, as a function of composition or temperature occur most frequently in that part of the periodic table. 4d and 5d compounds tend to be more metallic than the equivalent 3d compounds, since the radial extent of the d wavefunctions is larger than for 3d cations. However, there is no hard and fast rule that will exclude other possibilities; witness the metal-insulator transition in Ba(Pb/Bi)O$_3$ as a function of composition. Depending upon the oxidation state, transition elements from nickel to vanadium should be considered. If the d occupancy becomes small (say less than 4), the states at the Fermi level will be predominantly non-bonding and will tend to have somewhat less mixing than the unoccupied anti-bonding states at higher energy. This will make the largest difference to mechanisms based on phonon or exciton interactions, but may not be important to the RVB state at all.

Chlorides may also be good candidates, if metallic phases can be prepared. However, again the vast majority of chlorides are insulators. The p states of chlorine are close to those of oxygen, perhaps 0.5eV higher. Most of the same considerations applied above

to the case of nitrides would apply to the chlorides as
well.

Rather similar reasoning might lead one to
examine other anions as well, such as phosphide or even
hydride. The idea is to develop a set of unusual
features as a guide to future synthesis. If the cation-
anion mixing is believed to be important in your scheme
of imporant features, then for each anion one needs to
find the cations with electronic states close in energy
to the anion p states (or the s state in the case of
hydrides).

Summary

The unusual features of the new high temperature
superconductors have been outlined. It is suggested
that further synthetic studies, using some general
guidelines coupled with a few physical property
measurements, of new oxides as well as nitrides,
sulfides, and perhaps even chlorides may well lead to
more of these fascinating novel superconductors.

ACKNOWLEDGMENTS

This work was supported in part by the National Science
Foundation through the Materials Science Center at
Cornell University.

LITERATURE CITED

1. J. G. Bednorz and K. A. Müller Z. Phys. 1986, B64,
   189
2. R. J. Cava, R. B. van Dover, B. Batlogg, and E. A.
   Rietman Phys. Rev. Letts. 1987, 58, 408
3. M. K. Wu, J. R. Ashburn, C. J. Torng, P. H. Hor,

R. L. Meng, L. Gao, Z. J. Huang, Y. Q. Wang, C. W. Chu Phys. Rev. Letts. 1987, 58, 908

4. R. J. Cava, B. Batlogg, R. B. van Dover, D. W. Murphy, S. A. Sunshine, T. Siegrist, J. P. Remeika, E. A. Reitman, S. Zahurak, and G. P. Espinosa Phys. Rev. Letts. 1987, 58, 1676

5. M. A. Beno, L. Soderholm, D. W. Capone, D. G. Hinks, J. D. Jorgensen, I. K. Schuller, C. U. Segre, K. Z. Lang, J. D. Grace Appl. Phys. Letts. 1987, 51, 57

6. E. M. Engler, V. Y. Lee, A. I. Nazzal, R. B. Beyers, G. Lim, P. M. Grant, S. S. P. Parkin, M. L. Ramirez, J. E. Vasquez, and R. J. Savoy J. Amer. Chem. Soc. 109, 2848

7. D. B. Mitzi, A. F. Marshall, J. Z. Sum, D. J. Webb, M. R. Beasley, T. H. Geballe, and A. Kapitulnik Phys. Rev. B (to be published)

8. L. F. Mattheiss and D. R. Hamann Phys. Rev. 1983, B28, 4227

9. J. M. Tarascon, L. H. Greene, W. R. McKinnon, G. W. Hull, and T. H. Geballe Science 1987, 235, 1373

10. L. F. Matheiss Phys. Rev. Letts. 1987, 58, 1028

11. J. Yu, A. J. Freeman, and J. H. Xu Phys. Rev. Letts. 1987, 58, 1035

12. M. H. Whangbo, M. Evian, M. A. Beno, and J. M. William Inorg. Chem. 1987 26, 1831

13. see for example B. Raveau, C. Michel, and M. Hervieu in "Chemistry of High-Temperature Superconductors" eds. D. L. Nelson, M. S. Whittingham, and T. F. George, pub. ACS Symposium Series 351

14. L. F. Schneemeyer, J. V. Waszczak, S. M. Zahurak, R. B. van Dover, and T. Siegrist Mat. Res. Bull. 1987 (to be published)

15. A. Santoro, S. Miraglia, F. Beech, S. A. Sunshine,

D. W. Murphy, L. F. Schneemeyer, and J. V. Waszczak
Mat. Res. Bull. 1987, 22, 1007

16. T. M. Rice Z. fur Physik B 1987 (to be published)

17. P. Preovsek, T. M. Rice, and F. C. Zhang J. Physics
    C 1987, 20L

18. P. W. Anderson Science 1987, 235, 1196

19. T. Oguchi, H. Nishimori, and Y. Taguchi J. Phys.
    Soc. Japan 1986, 55, 323

20. V. J. Emery Phys. Rev. Letts. 1987, 58, 2794

21. Z. Kresin 1987 (to be published)

22. C. M. Varma, S. Schmidt-Rink, and E. Abrahams Solid
    State Commun. 1987, 62, 681

7

Precipitation of Superconductor Precursor Powders

BRUCE C. BUNKER, JAMES A. VOIGT, DANIEL H. DOUGHTY, DIANA L. LAMPPA, and KATHLEEN M. KIMBALL Sandia National Laboratories, Albuquerque, New Mexico

The most common synthetic method used to prepare the new high temperature superconducting ceramics such as $YBa_2Cu_3O_{7-x}$ (1,2,3) is to heat a mixture of ball-milled oxides and/or carbonates such as Y_2O_3, CuO, and $BaCO_3$ (1). The "mixed-oxide" synthesis has several potential problems, including: 1) Agglomerates in calcined powders are usually larger than 25 μm, producing ceramics which are highly porous and have poor mechanical properties. 2) Inadequate mixing of powders can lead to compositional inhomogenieties, inadequate stoichiometry control, the production of extraneous phases, and can introduce additional powder grinding steps which can influence the quality and reproducibility of the final product. Precipitation of powders from solution often provides a route for eliminating problems associated with mixed oxide processing (2). Through chemical precipitation routes, it is often possible to achieve much more intimate mixing of reagents prior to high temperature processing. The control of the stoichiometry and purity of chemically prepared powders can be higher than those obtained in mixed oxides. In most instances, finer powder particles can be produced via precipitation, which can promote sintering and lead to lower processing temperatures. We are investigating the preparation

This work performed at Sandia National Laboratories supported by the U.S. Department of Energy under contract number DE-ĸC04-76DP00789.

121

of superconductors using precipitated powders to see if we can prepare ceramics having improved physical and/or superconducting properties. Specifically, we have investigated how chemical processing routes influence the microstructures of superconducting ceramics and how the superconductivity is related to microstructural parameters.

In the precipitation methods we have investigated, a solution containing highly soluble metal salts such as chlorides or nitrates of yttrium, barium, and copper is mixed with a solution containing highly soluble salts of precipitating anions such as hydroxide or carbonate. When the solutions are mixed, an insoluble precipitate is formed. The precipitate is heated to form the desired superconducting phases. However, we find that if the precipitation process is not done carefully, the precipitate can actually be less homogeneous and pure than ball-milled mixed-oxides, leading to the synthesis of non-superconducting materials. The keys to successful precipitation are 1) the use of pH and anion concentration control to achieve the highest degree of supersaturation for all dissolved metal species, 2) the use of appropriate processing equipment to insure instantaneous and complete mixing of the soluble cation and soluble anion solutions, and 3) the use of counterions in both the cation and anion solutions which do not interfere with the precipitation process and which can be removed via high temperature processing.

The importance of pH and anion control in precipitation is illustrated in Fig. 1, which shows the calculated solubilities of Y^{3+}, Ba^{2+}, and Cu^{2+} in aqueous carbonate containing solutions as a function of pH. Solubilities were calculated using a code developed on Lotus 1-2-3[R]. Equilibrium concentrations of soluble hydroxide and carbonate and hydroxide species were calculated as a function of solution pH and carbonate concentrations from published solubility products and equilibrium formation constants (3,4) of the species of interest. The first point to note is that for successful precipitation, there must be a pH range in which all metals in the system are insoluble. When hydroxides alone are considered (the minimum set of complexes present in water), solubility diagrams reveal that $Ba(OH)_2$ is too soluble for complete precipitation. For

Superconductor Precipitation Calculation
$[CO_3]_{TOT} = 0.001$ M

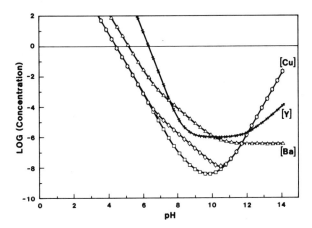

FIGURE 1 Calculated solubilities vs. solution pH for metal hydroxides and carbonates.

many other anion systems (including oxalates), at least one of the three metals (Y, Ba, or Cu) forms soluble complexes that reduce or eliminate the useful range for precipitation. However, as shown in Fig. 1, $BaCO_3$ is insoluble in basic solutions when excess carbonate (> 10^{-3}M) is present. By using a combination of carbonate and hydroxide, all three metals can be made to form insoluble precipitates between pH 9 and pH 12. The second point regarding pH control is that the desired pH for complete precipitation must be reached instantaneously, using special mixing to avoid severe concentration gradients in the solution. For example, the initial pH of metal cation solutions is below 4, in a regime where all three metal cations form soluble species. If the solution pH were to be gradually increased by pouring the anion solution into the cation solution, copper hydroxide would precipitate first, followed by barium carbonate and yttrium hydroxide. The resulting precipitate would not have a homogeneous chemical composition. We are able to achieve instantaneous mixing, producing the pH necessary for complete precipitation of homogeneous powder, by using a continuous precipitation system in which peristaltic pumps deliver separate cation and anion solutions into a turbulent mixing chamber. (However, homogeneous powders may not be required, since superconducting 1,2,3 samples can be prepared via a sequential hydroxide-carbonate precipitation route(5) similar to that described above.)

Although precipitates prepared using the solutions described above are insured of having the exact Y:Ba:Cu ratios present in the starting cation solution, the precipitates can be contaminated by the counterions used in both the cation and anion feed solutions. The contaminants are extremely difficult to remove via common washing procedures without partially redissolving the precipitates and changing product stoichiometries. For example, if potassium carbonate is dissolved as the carbonate source in the initial anion solution, the final product can contain as much as 1 wt% potassium, which interferes with subsequent sintering and produces unwanted sodium-containing phases in the final ceramic. Previous reports (5) of hydroxide-carbonate precipitation of 1,2,3 have not addressed this contamination problem. To avoid cation contamination, we use an anion solution which is a mixture of tetramethylammonium hydroxide and tetramethylammonium carbonate. The anion solution is prepared by bubbling CO_2 gas through a solution of tetramethylammonium hydroxide until the desired carbonate concentration and solution pH are reached. Although tetramethylammonium ions are incorporated in our precipitates, the contaminating cations are removed via pyrolysis during subsequent high temperature processing. A quaternary ammonium hydroxide must be used instead of the more common ammonium hydroxide because ammonia forms stable, highly soluble complexes with copper, preventing complete precipitation. As shown in Fig. 2, solution analyses confirm that we can achieve total precipitation of all dissolved Y, Ba, and Cu using the tetramethylammonium hydroxide/carbonate system if the solution pH is maintained between 9 and 12. There is good agreement between the solution analyses and calculated solubilities, especially considering that the

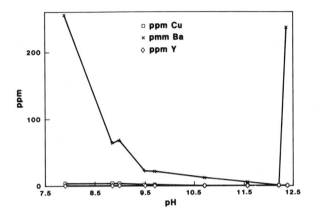

**All Metals Are Precipitated
Between pH 9 and 12**
Formula 123, Cl Salts, TMAOH + CO₂

FIGURE 2 Solution analyses of metal concentrations in filtrates
from 1,2,3 precipitates.

calculations model the three metals independently and do not account
for the possible formation of mixed-metal species.

The metal hydroxy-carbonate precipitate isolated from the
continuous precipitator after filtering, washing, and drying is an
amorphous, high surface area (100 m^2/gm) powder consisting of 1-3 μm
agglomerates. The precipitate must be heat treated to thermally
decompose the metal hydroxides, metal carbonates, and impurities to
convert them into the desired oxide $YBa_2Cu_3O_{7-x}$. Thermogravimetric
results (Fig. 3) and X-ray crystallography (Fig. 4) show that below
400°C, most of the hydroxides and carbonates have decomposed. By
650°C, the precipitate has been converted into an intimate mixture
of finely crystalline Y_2O_3, CuO, and $BaCO_3$. The desired 1,2,3 phase
is not formed until the calcining temperature exceeds 880°C, which
is the temperature at which $BaCO_3$ begins to decompose. By heating
at 920°C for appropriate times, it is possible to produce pure 1,2,3
which (by X-ray) is free of $BaCO_3$ and other impurity phases.
However, total calcination greatly lowers the surface area of the

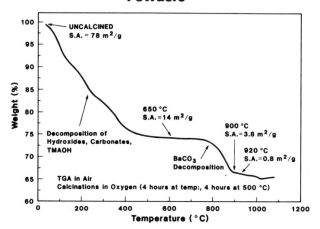

FIGURE 3   Thermogravimetric analysis of 1,2,3 precipitate.

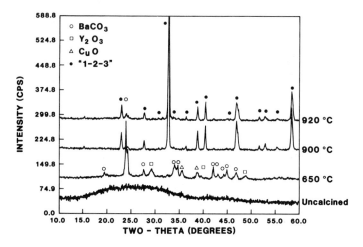

FIGURE 4   X-ray analysis of 1,2,3, precipitates vs. calcination temperature.

powder, producing interconnected 1 μm grains of 1,2,3 which are difficult to sinter into a dense ceramic material. Therefore, we have examined 1,2,3 materials prepared from partially calcined as well as totally calcined precipitates.

Sintering studies were conducted on different powder preparations using the same sintering schedules to determine the effect of precipitate chemistry and sintering conditions (atmospheres and temperatures) on the microstructure and superconductivity of 1,2,3 ceramics. Pressed pellets were heated in either air or oxygen at 950°C or 990°C for 4 hours followed by annealing in oxygen at 500°C for four hours. General conclusions of the sintering study are: 1) More sintering and grain growth occurs for samples heated in air instead of oxygen (Fig. 5), perhaps because the higher oxygen defect concentrations present in air-treated samples promote the formation of liquid phases above 950°C. 2) More densification and grain growth occurs for samples heated at 990°C instead of 950°C. 3) Calcination conditions have a strong influence on subsequent sintering. Powders calcined at 650°C sinter much better than those calcined at either 920°C (with large interconnected grains in the powder) or uncalcined powders, which sinter to porous structures due to excessive outgassing. 4) More sintering is observed for precipitates prepared from metal chloride solutions than from metal nitrates (Fig. 6). Although nitrate impurities are removed during calcining, chloride impurities remain in the samples, forming oxychloride phases which are liquid at high temperatures and promote sintering. The microstructures observed in the above study range from porous material (75% theoretical density) consisting of 1 μm equiaxed grains of 1,2,3 to >95% dense material consisting of 100 μm x 5 μm platelets of 1,2,3 material. Magnetic measurements (flux exclusion/flux expulsion) suggest that while all of the above materials are superconducting, the high density, large grained materials generally have a smaller volume fraction of 1,2,3 material which becomes superconducting during oxygen annealing than the porous, fine-grained microstructures. It appears that oxygen diffusion into the material during the annealing step is very slow in the dense 1,2,3 material. To conclude, we have thus far been

CHEM-PREP Powders Yield High
Density Materials (> 95%)

Chlorides pH = 10.5   650°C Calcine   950°C Sinter

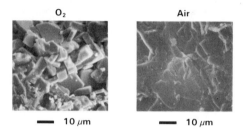

FIGURE 5   Scanning electron micrographs of 1,2,3 samples sintered in air and oxygen.

CHEM-PREP YBa$_2$Cu$_3$O$_{7-\Delta}$
Chlorides Sinter Better
Than Nitrates

pH = 10.5   650°C Calcine   950°C Sinter (O$_2$)

FIGURE 6   Scanning electron micrographs of 1,2,3 samples prepared from chloride or nitrate salt solutions.

able to use precipitated powders to produce high density 1,2,3 superconductors, or more porous samples with a larger volume of superconducting phase, but we have been unable to maximize both characteristics simultaneously.

ACKNOWLEDGMENTS

The authors wish to thank Gene Venturini for the magnetic characterization of our materials, Mike Eatough, Ralph Tissot, and Rich Lujan for extensive X-ray analyses, Bill Hammetter and Ed Binasiewicz for thermal analysis results, and Carol Ashley for surface area determinations.

REFERENCES

1. R. J. Cava, B. Batlogg, R. B. van Dover, D. W. Murphy, S.Sunshine, T. Siegrist, J. P. Remeika, E. A. Reitman, S. Zahurak, and G. P. Espinosa, Phys. Rev. Lett., 58, 1676 (1987).

2. K. D. Budd and D. A. Payne, Materials Research Society Symposia Proceedings, 32, 239 (1984).

3. C. F. Baes, Jr. and R. E. Mesmer, The Hydrolysis of Cations, John Wiley & Sons, New York, 1976.

4. J. Kragten, Atlas of Metal-Ligand Equilibria in Aqueous Solution, John Wiley & Sons, New York, 1978.

5. A. M. Kini, U. Geiser, H. C. I. Kao, K. D. Carlson, H. H. Wang, M. R. Monaghan, and J. M. Williams, Inorganic Chemistry, 26, 1836 (1987).

8

Synthesis and Characterization of High-T_c Superconductors in the $YBa_2(Cu_{1-x}Ni_x)_3O_{7-\delta}$ System

TENG-MING CHEN, JOSEPH F. BRINGLEY, and BRUCE A. AVERILL Department of Chemistry, University of Virginia, Charlottesville,
Virginia 22901

K. M. WONG and S. JOSEPH POON Department of Physics, University
of Virginia, Charlottesville, Virginia 22901

INTRODUCTION

Over the past year, the discovery of superconductivity above 90K
in a class of oxygen deficient cuprate perovskites (1,2) has
prompted an explosive growth of investigation on these materials.
The superconducting phase has been identified as a structurally
anisotropic layered pervoskite $LnBa_2Cu_3O_{7-x}$ (Ln = Y(2), La (3) and
rare earth (4), except Ce, Pr, Tb) with orthorhombic symmetry. In
particular, $YBa_2Cu_3O_{7-x}$ has been synthesized by various routes and
its crystal structure determined by both X-ray and neutron
diffraction methods (5,6). The mechanism and origin of superconductivity in $YBa_2Cu_3O_{7-\delta}$, however, remain ambiguous and un-
answered.

There are two crystallographically independent copper sites
in both the orthorhombic $YBa_2Cu_3O_{7-\delta}$ and the tetragonal
$YBa_2Cu_3O_{6+x}$ phases, namely, Cu(1) in the linear chains of corner-
shared CuO_4 square planes along the b-axis and Cu(2) in the layers
of corner-shared CuO_5 square pyramids parallel to the ab plane.
In addition, separate layers of Y^{3+} and Ba^{2+} cations fill the
spacing between the copper-oxygen layers and enforce the stability
of the structure. The fact that T_c remains almost unaffected upon
replacement of Y^{3+} by magnetic rare earth cations (with the

131

exception of Ce, Pr and Tb) indicates that superconductivity is
due to the low dimensional framework of copper and oxygen (3,4).

Recently, Xiao et al. (7) and Maeno et al. (8) reported an
investigation of the doping effect on superconductivity in
$YBa_2Cu_3O_{7-\delta}$ upon substitution of Cu by magnetic and nonmagnetic
elements. They found that the T_c's of doped $YBa_2(Cu_{1-x}M_x)_3O_{7-\delta}$
(M = Fe, Co, Ni, Cr, Mn, Zn, and Ga) were strongly suppressed by
doping with several atomic percent of the above elements.
However, no definitive conclusions regarding the mechanism of
superconductivity or structure-property relationships were drawn.
Tarascon et al. (9) have proposed that Ni^{2+} ions are in fact
disordered in both Cu sites and that the reduction in T_c in these
oxides is due to disorder, rather than to a magnetic effect.   In
addition, no structural transformation in $YBa_2Cu_{3-x}Ni_xO_{7-\delta}$ (x $\leq$
0.50) was observed.

To extend the previously reported results, we have further
examined the $YBa_2(Cu_{1-x}Ni_x)_3O_{7-\delta}$ system. We wished to determine
the solubility limit of Ni in the parent $YBa_2Cu_3O_{7-\delta}$ phase and the
effect of doping on the oxygen stoichiometry. We have investigated the shielding and Meissner effect to determine the volume
fraction of the superconducting phase present in the bulk materials. Magnetic susceptibility measurements above T_c were also
carried out to provide additional information on the electronic
density of states and the valence states of the Cu ions.   The
determination of the upper critical field of $YBa_2(Cu_{1-x}Ni_x)_3O_{7-\delta}$
is also reported.

METHODS

Synthesis

The formation of $YBa_2(Cu_{1-x}Ni_x)_3O_{7-\delta}$ was systematically investigated by substituting nickel for copper. Mixtures of NiO, Y_2O_3,
$BaCO_3$ and Cu or CuO powders in appropriate proportions were
thoroughly mixed, ground and then calcined in air at 900-950°C for
16 hours in alumina crucibles. The black products were then
pulverized and cold-pressed into pellets (13mm o.d. x 3mm), which

were then fired once under similar conditions.

Oxygen annealing on the pellets was carried out at 900-950°C for 20 hours to sinter the materials. The temperature was then lowered to 500°C for 16 hours and the samples were finally cooled to ambient temperature over a six-hour period to obtain the orthohombic phase. Samples prepared from fine copper powder (200 mesh) were found to be superior to those made from CuO based on the purity of the 123 phase as determined by X-ray powder diffraction.

We find that up to 20% of nickel (i.e. x = 0.20) can be doped or dissolved into the copper sites in the lattice to form $YBa_2(Cu_{1-x}Ni_x)_3O_{7-\delta}$, which produced a systematic shift in the X-ray powder patterns. An unknown phase plus green NiO started to separate from the rest of the products when $x \geq 0.30$, indicating the maximum solubility of nickel had been exceeded.

RESULTS AND DISCUSSION

Powder X-ray diffraction measurements

All of the samples of $YBa_2(Cu_{1-x}Ni_x)_3O_{7-\delta}$ with $x \leq 0.30$ prepared from copper powder were characterized by powder X-ray diffraction and were found to consist of an almost pure single phase for $x \leq 0.20$. The cell parameters were obtained by indexing the powder patterns using a least-square program LATT (10), using the superconducting orthorhombic phase as a starting point.

The lattice parameters of $YBa_2(Cu_{1-x}Ni_x)_3O_{7-\delta}$ ($0 \leq x \leq 0.30$) are plotted against the nominal composition of nickel, x, in Fig. 1. A systematic variation of the individual dimension is observed, resulting in a decrease in unit cell volume with increasing nickel concentration up to x = 0.20. A similar trend was also reported by Tarascon et al. (9) for $x \leq 0.17$. The cell contraction may be attributed to the smaller size of Ni^{2+} (octahedral radius 0.70Å) vs. that of Cu^{2+} (0.73 Å) (11). We also noted that the formation of $YBa_2(Cu_{1-x}Ni_x)_3O_{7-\delta}$ does not induce a structural transformation, such as is observed in the doped $YBa_2(Cu_{1-x}M_x)_3O_{7-\delta}$ (M = Fe, Co and Ga) phases (8).

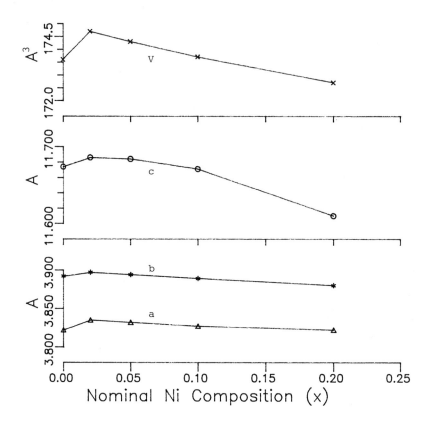

FIGURE 1. Variation of the orthorhombic lattice parameters a, b, c and V as a function of x for Ni-doped $YBa_2(Cu_{1-x}Ni_x)_3O_{7-\delta}$ phases.

Electrical resistivity measurements

The temperature dependence of the resistivity of $YBa_2(Cu_{1-x}Ni_x)_3O_{7-\delta}$ ($0 \leq x \leq 0.20$) is represented in Fig. 2. Standard four-lead conductivity measurements on all sintered pellet samples were carried out from 300 K down to 4.2 K. Above the superconducting transition temperature (T_c), all of the samples exhibited poor metallic behavior with resistivities ranging from 8.8 mΩ-cm for x = 0.02 to 82 mΩ-cm for x = 0.20 at room temperature.

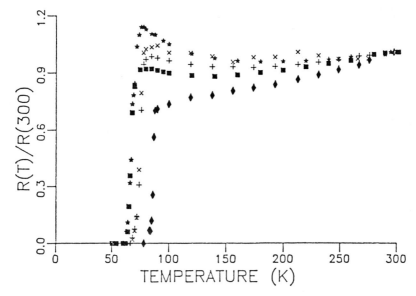

FIGURE 2. Temperature dependence of the normalized resistivity of YBa$_2$(Cu$_{1-x}$Ni$_x$)$_3$O$_{7-\delta}$ with x = 0.02 (♦), 0.05 (■), 0.10 (+), 0.20 (x) and 0.30 (⁎).

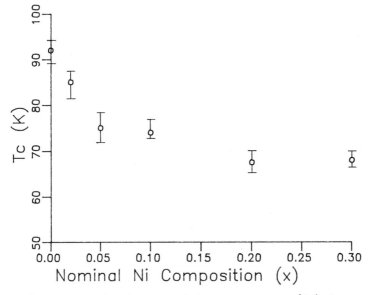

FIGURE 3. Superconducting transition temperature (T$_c$) for YBa$_2$(Cu$_{1-x}$Ni$_x$)$_3$O$_{7-\delta}$ with 0 ≤ x ≤ 0.30. The temperatures for 10% and 90% drops in resistivity are indicated by vertical bars.

The superconducting temperature for $0 \leq x \leq 0.30$ as determined by the resistivity measurements are summarized in Fig. 3 for
various nominal nickel compositions, x, with vertical bars
indicating the temperatures at which 10% and 90% drops in resistivity occur. Doping the $YBa_2Cu_3O_{7-\delta}$ phase with the magnetic Ni^{2+}
ion induces a gradual reduction in T_c, but does not destroy
superconductivity for the phases with $x \leq 0.30$ investigated. It
is interesting to note that the conduction electrons appear to
become weakly localized as the concentration of nickel increases
before superconductivity sets in. Since the same trend in T_c and
resistivity is observed on both quenched phases of $YBa_2Cu_3O_{7-\delta}$
(12,13) and $YBa_2(Cu_{1-x}Ni_x)_3O_{7-\delta}$, we postulate that nickel doping
may be equivalent to decreasing oxygen stoichiometry. Thus the
ratio of Cu^{3+} to Cu^{2+} in the doped $YBa_2(Cu_{1-x}Ni_x)_3O_{7-\delta}$ phases is
significantly reduced. The determination of accurate oxygen
contents in nickel-doped $YBa_2Cu_3O_{7-\delta}$ is currently in progress.

Magnetic studies

In order to study the homogeneity of the superconducting samples
and to select high-quality samples for more detailed studies, we
have measured the shielding and Meissner effect. The d.c.
magnetic susceptibility was measured from 300K to 4K using a SHE
SQUID susceptometer in a magnetic field of 100 G (Meissner
effect). The calibration of the instrument and measurement
techniques have been described previously (14,15). In addition,
$YBa_2(Cu_{1-x}Ni_x)_3O_{7-\delta}$ phases with x = 0.02 and 0.05 were warmed up
in a 100G field after cooling in a zero field to measure the
shielding effect. Results are shown in Figure 4.

The low Meissner effect of 12 and 8.5% by volume for the
samples with x = 0.02 and 0.05, respectively, appears to be
typical in this type of material (16). For samples with x = 0.05,
the transition observed in Fig. 4 was broader than that for a
sample with x = 0.02. The gram susceptibility for both compositions reached a constant value only below ca. 20 K. In addition
to grain size effects, we suspect that the low Meissner effects

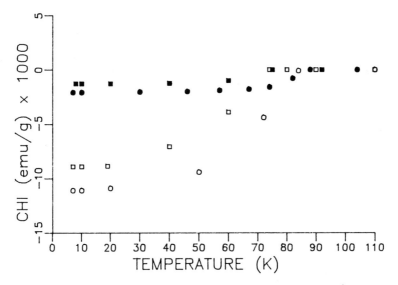

FIGURE 4. Gram magnetic susceptibility vs. temperature (Meissner effect (solid symbols) and shielding (open symbols)) for samples of nominal composition YBa$_2$(Cu$_{1-x}$Ni$_x$)$_3$O$_{7-\delta}$ with x = 0.02 (circles) and 0.05 (squares) in a magnetic field of 100 G.

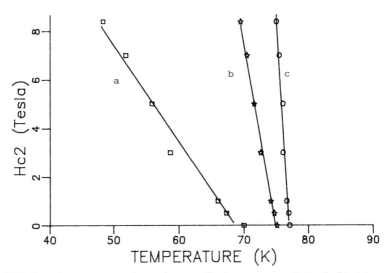

FIGURE 5. Temperature dependence of the upper critical field (H$_{c2}$) for the nominal superconducting phase YBa$_2$(Cu$_{0.95}$Ni$_{0.05}$)$_3$O$_{7-\delta}$ near T$_c$ for 90% (a), 50% (b), and 10% (c) resistivity drops.

are due to the fact that the actual amount of superconducting
phase in the samples may be lower than that expected from the X-
ray diffraction data.

The upper critical fields $H_{c2}(T)$ were determined with a
standard four-probe technique in a d.c. magnetic field up to 9
Tesla. Silver paint was applied to attach the leads on a cold-
pressed chunk of $YBa_2(Cu_{0.95}Ni_{0.05})_3O_{7-\delta}$ with x = 0.05. The data
for $H_{c2}(T)$ near T_c for the superconducting phase are shown in
Fig. 5. The $-\frac{dH_{c2}}{dT}$ T_c values were determined to be 4.1, 1.5 and
0.39 T/K for different $H_{c2}(T)$'s, where the resistivity drops to a
fraction of 90, 50 and 10% of its extrapolated normal-state value
at that field. These values are somewhat greater than that
observed in the undoped $YBa_2Cu_3O_{7-\delta}$ phase.

CONCLUSIONS

We have prepared a series of Ni-doped $YBa_2Cu_3O_{7-\delta}$ phases with x $\leq$
0.20. Systematic decreases in cell dimensions and the unit cell
volume in the nickel-doped phases indicate the formation of
$YBa_2(Cu_{1-x}Ni_x)_3O_{7-\delta}$ with x up to 0.20. Oxygen annealing after
initial heat treatment is crucial for the occurrence of superconductivity, as in the parent pure copper phase.

The superconducting transition temperature, T_c, decreases
with increasing concentration of doped magnetic Ni^{2+} ions, but
superconductivity is not abolished for x $\leq$ 0.20. The low Meissner
effect (approximately 10% by volume) and the diamagnetic shielding
for 2 and 5% Ni-doped $YBa_2Cu_3O_{7-\delta}$ samples may be due to the effect
of grain size and grain morphology.

Based on the above results, we conclude that the suppression
of T_c in $YBa_2(Cu_{1-x}Ni_x)_3O_{7-\delta}$ (0 $\leq$ x $\leq$ 0.20) has both intrinsic as
well as extrinsic origins. Thus, it may be attributed to a
combination of structural (e.g., type of Cu site replaced),
electronic (e.g., the presence of Ni^{2+}/Ni^{3+}), and magnetic

contributions. Further studies will be necessary to assess the importance of each of these.

ACKNOWLEDGMENTS

This research was supported in part by the National Science Foundation, Solid State Chemistry Program, Grant DMR-8313252 (B. A. A., T.-M. C., J. F. B.) and the Jeffress Memorial Trust (K. M. W.).

REFERENCES

1. Bednorz, J. G.; Müller, K. A., Z. Phys B: Condens. Matter, **64** 189, (1986).

2. Chu, C. W.; Hor, P. H.; Meng, R. L.; Geo, L.; Huang, Z. J.; Wang, Y. Q., Phys. Rev. Lett., **58**, 408 (1987).

3. Chu, C. W., Proc. Natl. Acad. Sci. USA, **84**, 4681 (1987).

4. Engler, E. M.; Lee, V. Y.; Nazzul, A. I.; Beyers, R. B.; Lim, G.; Grant, P. M.; Parkin, S. S. P.; Ramirez, M. L.; Vazquez, J. E. and Savoy, R. J., J. Am. Chem. Soc., **109**, 2848 (1987).

5. Siegrist, T.; Sunshine, S.; Murphy, D. W.; Cava, R. J. and Zahurak, S. M., Phys. Rev. B., **35**, 7137 (1987).

6. David, W. I. F., Nature, **327**, 310 (1987).

7. Xiao, G.; Streitz, F. H.; Garrin, A.; Du, Y. W.; and Chien, C. L., Phys. Rev. B., **35**, 8782 (1987).

8. Maeno, Y.; Tomita, T.; Kyogku, M.; Awazi, S.; Aoki, Y.; Hoshino, M.; Minami, A. and Fujita, T., Nature, **328**, 512 (1987).

9. Tarascon, J. M.; Barboux, P.; Bagley, B. G.; Greene, L. H.; McKinnon and Huyll, G. W., in "Chemistry of High-Temperature Superconductors"; Nelson, D. L.; Wittingham, M. S. and George, T. F., Ed.; ACS Symposium Series 351; American Chemical Society: Washington, D.C., 1987, Chapter 20.

10. Takusagawa, F., Ames Laboratory, Iowa State University, unpublished research, 1981.

11. Shannon, R. D., <u>Acta Crystallogr.</u>, **A32**, 7511 (1965).

12. Schuller, I. K.; Beno, M. A.; Capone II, D. W.; Bruynseraede,
     Y.; Egre, S. U.; Hinks, D. G.; Locquet, J.-P.; Soderholm, L.;
     Segre, S. U.; Zhang, K., <u>Solid State Commun.</u>, **63**, 385 (1987).

13. van den Berg, J., et al. <u>Europhys. Lett.</u> (submitted).

14. O'Connor, C. J.; Deaver, B. S., Jr.; Sinn, E., <u>J. Chem.
     Phys.</u>, **70**, 516 (1979).

15. O'Connor, C. J.; Sinn, E.; Cukauskas, E. J.; Deaver, B. S.,
     Jr., <u>Inorg. Chim. Acta</u>, **32**, 39, (1979).

16. Dinger, T. R.; Worthington, T. K.; Gallagher, W. J. and
     Sandstrom, R. L., <u>Phys. Rev. Lett.</u>, **58**, 2687 (1987).

9

Preparation and Physical Properties of Oxyfluorides in the Ln-Ba-Cu System (Ln = Y, La)

JOSEPH F. BRINGLEY, TENG-MING CHEN, and BRUCE A. AVERILL Department of Chemistry, University of Virginia, Charlottesville, Virginia 22901

K. M. WONG and S. JOSEPH POON Department of Physics, University of Virginia, Charlottesville, Virginia 22901

INTRODUCTION

The report by Bednorz and Müller (1) of possible high T_c superconductivity in the La-Ba-Cu system sparked an intense scientific effort toward identification of new high T_c materials. To date, two classes of materials defined by the prototypical $La_{1.8}Sr_{0.2}CuO_{4-\delta}$ ($T_c \approx 35K$) and $YBa_2Cu_3O_{7-\delta}$ ($T_c \approx 93K$) have been discovered (2-6). Recent studies have shown that the structure and physical properties are strongly affected by the anion stoichiometry and local structure of the Cu chains in $YBa_2Cu_3O_{7-\delta}$ (7,8). We have chosen to investigate oxyfluorides in these systems because of fluoride's demonstrated ability to substitute for oxide (9) and its ability to support a magnetic coupling (though somewhat weaker than oxide) between metal atoms, and to further probe the role of the anion in determining the physical properties of these materials.

We have investigated the synthesis of Ln-Ba-Cu oxyfluorides by a number of methods, including direct solid state reaction, reaction with KF flux, and annealing under F_2 gas. We report herein the results of these studies.

EXPERIMENTAL

Starting materials (La_2O_3, Y_2O_3, $BaCO_3$, CuO or Cu powder, BaF_2,

CuF_2, YF_3, and LaF_3) were obtained as high purity ($\geq$ 99.99%) materials where possible. Lanthanide oxyfluorides (LnOF) were prepared as described (10). Reactions were carried out in Pt or alumina crucibles under air, O_2 or dry Ar atmospheres. All materials were carefully checked for purity by powder X-ray diffraction. Oxygen-deficient $YBa_2Cu_3O_{6.2}$ was obtained by annealing a single phase air-annealed sample under Ar at $\approx$900°C for 12 h. The oxygen content (prior to fluorination) was determined by hydrogenation using a 50:50 H_2/Ar mixture at 1000°C. Fluorination reactions were carried out in a home-constructed Cu line and vessel equipped with an HF scrubber and precision flow meter. Fluorine uptake was estimated gravimetrically by careful weighing ($\pm$ 0.05 mg) before and after fluorination. Powder X-ray diffraction data were obtained by the Guinier technique using an Enraf-Nonius diffractometer equipped with a Si monochrometer to give clean Cu $K_{\alpha1}$ radiation. Temperature-dependent resistivity data were measured by the four-probe technique, and Meissner effect data were obtained on a SHE SQUID magnetometer.

RESULTS

Initial attempts to prepare Ln-Ba-Cu oxyfluorides were carried out by direct solid state reaction of the appropriate metal oxides and metal fluorides or oxyfluorides. The fluorides investigated included CuF_2, BaF_2, LnF_3 and LnOF (Ln = Y or La). No evidence suggesting the formation of $LnBa_2Cu_3O_{7-x}F_x$ (0.1 $\leq$ x $\leq$ 2.0) or $La_{2-x}Ba_xCuO_3F$ could be inferred from powder diffraction data, but in each case BaF_2 was present in the product. The electrical properties of the products were screened and, for Ln = Y, showed sharp resistivity drops between 90-100K and zero resistance near 90K. The synthesis reported by Ovshinsky et al. (11) was also carefully repeated under a variety of conditions; in each case multiphase products containing BaF_2 with superconducting transition temperatures near 90K were obtained. The products afforded by direct methods are indicated in Eq. 1 below.

$$Ln_2O_3 + 4BaCO_3 \longrightarrow x \, BaF_2 + Y_2BaCuO_5 \quad (\underline{1})$$
(minor)

$$+ (6-x)CuO + xCuF_2 \qquad + YBa_2Cu_3O_{7-\delta} + CuO$$
(major)

$$(0.1 \leq x \leq 2.0)$$

Reaction of deoxygenated $YBa_2Cu_3O_{7-\delta}$ ($\delta \approx 1.0$) with a KF flux was also investigated as a route to oxyfluorides. Reactions were carried out above the melting point of KF ($\geq 860°C$) and yielded only decomposition products with traces of the orthorhombic 123 phase.

Ln-Ba-Cu oxyfluorides were successfully prepared by reaction of $YBa_2Cu_3O_x$ ($6.2 \leq x \leq 6.9$) with pure F_2 gas at moderate temperatures ($130 - 175°C$, $2 - 4d$). Reaction temperatures above $200°C$ resulted in slow decomposition of the materials to amorphous products, presumably BaF_2, YF_3 or YOF, and CuF_2 or Ba_2CuF_6. Reaction of tetragonal, deoxygenated $YBa_2CuO_{6.2}$ with pure F_2 gas ($170°C$, $4d$), yielded a material analyzed as $YBa_2Cu_3O_{6.2}F_{0.63}$. Analysis of the powder pattern revealed a transition to orthorhombic symmetry upon fluorination. X-ray powder pattern data of the starting material and product are shown in Figure 1. The orthorhombic distortion can clearly be seen as the 013/103, 020/200, and 123/213 lines are split into separate components. As shown in Table 1, the cell volume decreases significantly upon fluorination, but the effect is not as dramatic as when the vacancies are filled by oxide. This can be understood in terms of fluoride's decreased polarizability and lower charge, resulting in increased ionic character and longer metal-anion bonds (9).

A range of anion compositions in $YBa_2Cu_3O_xF_y$ could be obtained depending upon experimental conditions; a partial list of characterized samples is given in Table 2. All materials were single phase by X-ray powder diffraction, except where noted. The materials $YBa_2Cu_3O_xF_y$ ($6.2 \leq x \leq 6.9$; $0.1 \leq y \leq 0.63$) show semiconducting behavior down to liquid He temperature for $y > 0.5$ (Figure 2). The material $YBa_2Cu_3O_{6.2}F_{0.25}$ lies in a two-phase

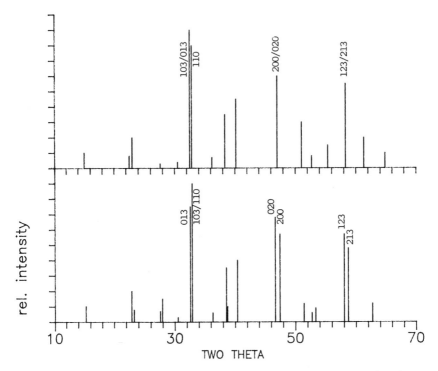

FIGURE 1. X-ray powder diffraction data for $YBa_2Cu_3O_{6.2}$ (top) and $YBa_2Cu_3O_{6.2}F_{0.63}$ (bottom).

TABLE 1. Unit Cell Parameters for $YBa_2Cu_3O_{7-x}$ (x = 0, 0.8) and $YBa_2Cu_3O_{6.2}F_{0.63}$

	a(Å)	b(Å)	c(Å)	V(Å³)
x ≈ 0.8	3.858	3.870	11.748	175.4
x ≈ 0.0	3.822(1)	3.892(1)	11.674(3)	173.65(5)
$YBa_2Cu_3O_{6.2}F_{0.63}$	3.824(2)	3.894(2)	11.685(7)	174.0(2)

TABLE 2. New Oxyfluorides Prepared by Direct Fluorination

Composition	Temp. (time)[a]	Cell Parameters (Å)	Properties
$YBa_2Cu_3O_{6.2}F_{0.63}$	170°C (4d)	a = 3.824(2) b = 3.894(2) c = 11.685(7)	ρ_{RT} = 1.3 Ωcm $\rho(77K)/\rho_{RT}$ = 5.46 $\rho(4K)/\rho_{RT}$ = 12.1
$YBa_2Cu_3O_{6.2}F_{0.54}$	170°C (3d)	a = 3.827 b = 3.884 c = 11.673	ρ_{RT} = 280 Ωcm $\rho(77K)/\rho_{RT}$ = 46
$YBa_2Cu_3O_{6.2}F_{0.25}$	110°C (2d)	tetragonal + orthorhombic phases present	ρ_{RT} = 44 mΩcm T_c^{mid} = 85K, $T_{co} \approx$ 50K χ_V (Meissner) = 2%
$YBa_2Cu_3O_{6.9}F_{0.11}$	150°C (3d)	-	ρ_{RT} = 4.8 mΩcm T_c^{mid} = 90K, T_{co} = 89K χ_V (Meissner) = 17%
$La_{1.8}Ba_{0.2}CuO_{3.9}$	-	a = 3.792 c = 13.274	ρ_{RT} = 4.2 mΩcm T_c^{mid} = 30K, T_{co} = 25K
$La_{1.8}Ba_{0.2}CuO_{3.9}F_{0.11}$	150°C (3d)	a = 3.791 c = 13.264	ρ_{RT} = 22 mΩcm $\rho(77K)/\rho_{RT}$ = 0.616 T_c^{mid} = 25K, $T_c \approx$ 10K

[a] Conditions for F_2 annealing.

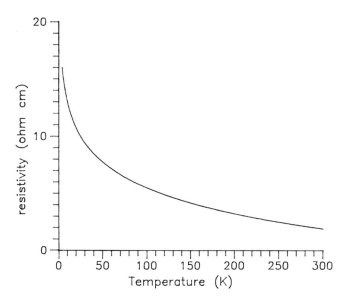

FIGURE 2. Resistivity data for $Ba_2YCu_3O_{6.2}F_{0.63}$ as a function of temperature.

region exhibiting both tetragonal and orthorhombic symmetry. A
broad superconducting transition is observed with zero resistance
at 50K. However, a 2% Meissner effect suggests the superconductivity is due only to a minority phase, possibly formed along the
grain boundaries.   Fluorination of O_2-annealed $YBa_2Cu_3O_{6.9}$ results
in uptake of fluoride giving $YBa_2Cu_3O_{6.9}F_{0.11}$. Fluoride incorporation into samples such as this, with high oxygen content, results
in insignificant changes in cell parameters and has little effect
upon the superconducting transition temperature.

The electrical resistivities of $La_{1.8}Ba_{0.2}CuO_{3.9}$ (214
structure) before and after fluorination are plotted vs. temperature in Figure 3. Upon incorporation of 0.11 moles of fluoride T_c
decreases from 25K for $La_{1.8}Ba_{0.2}CuO_{3.9}$ to 10K for
$La_{1.8}Ba_{0.2}CuO_{3.9}F_{0.11}$.

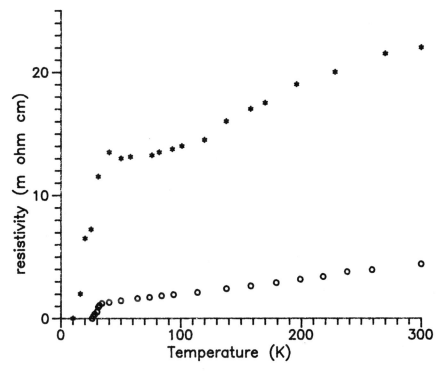

FIGURE 3. Resistivity data plotted for $La_{1.8}Ba_{0.2}Cu_{3.9}$ (open circles) and $La_{1.8}Ba_{0.2}CuO_{3.9}F_{0.11}$ (stars) as a function of temperature.

DISCUSSION

The ability of fluoride to substitute for oxide has been well documented (9,12-18). The substitution can occur in materials where mixed valence states are available or where charge compensation can be achieved by appropriate cationic replacements. The high T_c materials thus pose excellent candidates for fluoride substitution since either of these criteria can in principle be fulfilled. Indeed a number of perovskite oxyfluorides are known (e.g., $Sr_2FeO_{3+x}F_{1-x}$ (14), $Ca_2MnO_{4-x}F_x$ (15), K_2NbO_3F (16a)) with the K_2NiF_4 type structure. The electrical properties of oxyfluorides range from insulating (e.g., Sr_2FeO_3F (16b)) to metallic (e.g., $V_2O_{4-x}F_x$ (17)) to superconducting (e.g., $A_xWO_{3-x}F_x$; A = Li, K, Rb, Cs (18)).

A key feature of the Ln-Ba-Cu oxide superconductors is considerable mixing of the Cu 3d and O 2p bonds (19). Incorporation of fluoride into metal oxides introduces an admixture of partly covalent (M-O) and mostly ionic (M-F) bonding within the structure (9). Although fluoride is slightly smaller than oxide (ionic radius 1.31 Å for F^- vs. 1.38 Å for O^{2-}), its lower anionic charge and higher electronegativity result in a reduced tendency to π-bond and in slightly longer M-F bonds than for M-O. This significantly reduces metal-fluoride orbital mixing when compared to that of metal oxides. Indeed, the fluoride orbitals are 2-3 electron volts lower in energy than those of oxygen (19). Thus, if covalency between metal d and anion p-bands is important for high T_c superconductivity, fluorides would certainly be poor candidates. On the other hand, if low dimensionality is favorable for high T_c, then the incorporation of fluorides (if ordered) may enhance T_c by increasing the anisotropy of the materials.

In light of our results, it appears that direct methods are not suitable for the preparation of oxyfluorides related to the high T_c Ln-Ba-Cu-O superconductors. The thermal stability of BaF_2 at the temperatures necessary to form the 214 and 123 phases appears to constitute the major barrier. In contrast to Ovshinsky et al. (11), we have not found T_c > 90K in these systems.

Upon annealing in F_2 gas, deoxygenated, nearly tetragonal $YBa_2Cu_3O_{6.2}$ undergoes a transition to orthorhombic symmetry (Figure 2). This is the same effect as is observed when annealing under pure O_2 and is consistent with a filling of the anion vacancies along the b-axis. The increase in cell volume for $YBa_2Cu_3O_{6.2}F_{0.63}$ when compared with $YBa_2Cu_3O_{6.9}$ (3) is consistent with longer M-F bonds and has been observed elsewhere (17,20). The physical properties of the materials suggest semiconducting behavior at relatively high F^-/O^{2-} ratios (i.e., > 0.5/6.2),

although the observation of superconductivity in these materials
is not impossible if microdomains of the high T_c phase can form to
create a percolative path. For small F^-/O^{2-} ratios (i.e., $\approx$
0.1/6.9), T_c remains largely unaffected and the materials display
bulk superconductivity.

The K_2NiF_4 structure seems more amenable to fluoride substitution, as superconductivity is not completely abolished in
fluorinated samples of $La_{1.8}(Ba,Sr)_{0.2}CuO_{4-\delta}$ ($\delta \leq 0.2$) However,
T_c is diminished in $La_{1.8}Ba_{0.2}CuO_{3.9}F_{0.11}$ ($T_c = 10K$) vs.
$La_{1.8}Ba_{0.2}CuO_{3.9}$ ($T_c = 25K$). The greater tolerance for fluoride
substitution may be due to the greater two-dimensional character
of the K_2NiF_4 structure as opposed to the $YBa_2Cu_3O_7$ structure.

SUMMARY

We have developed a method for preparing Ln-Ba-Cu oxyfluorides
with the K_2NiF_4 and $YBa_2Cu_3O_7$ phases. Tetragonal $YBa_2Cu_3O_{7-\delta}$ ($\delta \leq$
0.8) undergoes an orthorhombic transition upon annealing under
pure F_2 gas at 170°C, implying a filling of the anion vacancies
along the b axis by fluoride. In general, incorporation of
fluoride produces semiconductors for $YBa_2Cu_3O_7$ and diminished T_c
for K_2NiF_4 phases. These results emphasize the importance of Cu-
anion covalency and the CuO_2 chains as key factors for high T_c
superconductivity.

ACKNOWLEDGMENTS

This research was supported in part by the National Science
Foundation, Solid State Chemistry Program, Grant DMR-8313252 (B.
A. A., T.-M. C., J. F. B.) and Jeffress Memorial Trust (K. M. W.).

REFERENCES

1. Bednorz, J. G., and Müller, K. A., Z. Phys., B64, 189 (1986).

2. Wu, M. K.; Ashburn, J. R.; Torng, C. J.; Hor, P. H.; Meng, R.
    L.; Gao, L.; Huang, Z. J.; Wang, Y. Q., and Chu, C. W., Phys.
    Rev. Lett., **58**, 908 (1987).

3. Cava, R. J.; Battlogg, B.; van Dover, R. B.; Murphy, D. W.;
    Sunshine, S.; Siegrist, T.; Remeika, J. P.; Reitman, E. A.;
    Zahurak, S., and Espinosa, G. P., Phys. Rev. Lett., **58**, 1676
    (1987).

4. Tarascon, J. M.; Green, L. H.; McKinnon, W. R.; Hull, G. W.,
    Phys. Rev. B, **35**, 7115 (1987).

5. Grant, P. M.; Beyers, R. B.; Engler, E. M.; Lim, G.; Parkin,
    S. S.; Ramirez, M. L.; Lee, V. Y.; Nazzal, A.; Vasquez, J.
    E., and Savoy, R. J., Phys. Rev. B., **35**, 7242 (1987).

6. Chu, C. W.; Hor, P. H.; Meng, R. L.; Gao, L.; Huang, Z. J.;
    and Wang, Y. Q., Phys. Rev. Lett., **58**, 405 (1987).

7. Murphy, D. W.; Sunshine, S. A.; Gallagher, P. K.; O'Bryan, H.
    M.; Cava, R. J.; Batlogg, B.; van Dover, R. B.; Schneemeyer,
    L. F., and Zahurak, S. M., in Chemistry of High-Temperature
    Superconductors, D. L. Nelson, M. S. Whittingham and T. F.
    George, eds., ACS Symposium Series, New Orleans, LA (1987),
    pp. 181-197.

8. Johnston, D. C.; Jacobson, A. J.; Newsman, J. M.;
    Lewandowski, J. T.; Goshorn, D. P.; Xie, D., and Yelon, W.
    B., ibid (1987), pp. 136-151.

9. (a) Bartlett, N. in Inorganic Solid Fluorides, Hagenmuller,
    P. ed., Academic Press, London, Eng. (1985), pp. xiii-xiv.
    (b) Hagenmuller, P., ibid (1985), p. xv. (c) Chamberland, B.
    L., ibid (1985), pp. 205-258.

10. Shinn, D. B., and Eick, H. A., Inorg. Chem., **8**, 232 (1969).

11. Ovshinsky, S. R.; Young, R. T., Allred, D. D., DeMaggio, G.,
    and van der Leeden, G. A., Phys. Rev. Lett., **58**, 2579 (1987).

12. Sleight, A. W., Inorg. Chem., **8**, 1764 (1969).

13. Robbins, M.; Pierce, R. D., and Wolfe, R., J. Phys. Chem.
    Sol., **32**, 1789 (1971).

14. Menil, F.; Kinomura, N.; Fournes, L.; Portier, J., and Hagenmuller, D., Phys. Stat. Sol. (a), **64**, 261 (1981).

15. (a) Le Flem, G.; Colmet, R.; Claverie, J.; Hagenmuller, P., and Georges, R., J. Phys. Chem. Sol., **41**, 55 (1980).

16. (a) Galasso, F., and Darby, W., J. Phys. Chem., **66**, 1318 (1962). (b) Galasso, F., and Darby, W., J. Phys. Chem., **67**, 1451 (1963).

17. Chamberland, B. L., Mat. Res. Bull., **6**, 425 (1971).

18. Hubble, F. F.; Gulick, J. M., and Moulton, W. G., J. Phys. Chem. Sol., **32**, 2345 (1971).

19. Disalvo, F. J., in Chemistry of High-Temperature Superconductors, D. L. Nelson, M. S. Whittingham, and T. F. George, eds., ACS Symposium Series, New Orleans, LA (1987).

20. Chamberland, B. L.; Frederick, C. G., and Gillson, J. L., J. Sol. St. Chem., **6**, 561 (1973).

10

Oxalate Precipitation Methods for Preparing the Yttrium-Barium-Copper Superconducting Compound

RONALD J. CLARK   Department of Chemistry, Florida State University
Tallahassee, Florida 32306

WILLIAM J. WALLACE   Visiting Faculty Scholar, Department of
Chemistry, Muskingum College, New Concord, Ohio 43762

JENNIFER A. LEUPIN   Undergraduate Summer Participant, Department of
Chemistry, University of West Florida, Pensacola, Florida 32504

The current excitement over the new ceramic materials that become superconductors above 78K hardly needs to be discussed. Less
than a year after the first materials really came to the attention
of the scientific community, there was an ACS meeting in New Orleans
in which three full days of papers were presented. Most of the
papers concerned the characterization of the 40 and 90 K materials
or attempts to modify them, but only a few concerned their
synthesis.

The preparation of the basic material is a relatively simple
matter. One takes appropriate quantities of the oxides and/or
carbonates which are mixed by grinding, followed by an initial
firing, a further regrinding, pelletization, and a final heat
treatment in oxygen. Generally, only the final step has critical
aspects to it. There are several alternate methods of preparation
such as solutions techniques which include the carbonate and the
citrate procedures.[1] Due to chemical problems, organometallic
solgel methods[2] have not yet been successful owing to a lack of
appropriate starting materials. The oxalate method, originally
successful for the strontium lanthanum material[3] has been
criticized[1] for use with the yttrium-barium materials. Although

there are potential problems in the use of that method, we have
chosen to start a systematic investigation of the procedure. A
potential advantage of the technique comes from the possibility of
producing a much more finely divided and intimately mixed product
than that which results by the standard procedure. This, in turn,
might help overcome problems such a bulk density and critical
current capacity.

    If one assumes that the three metal oxalate compounds behave
independently, there are a number of reaction equilibria which must
be considered. In addition to the three Ksp values, one needs to
be concerned with pK values of oxalic acid, with complexation
(solubilization) of the metals, with hydroxide formation, and also
possibly with the solubility of the neutral salt. Calculations
using the available literature data clearly indicate the most
severe problems concern the solubility of the barium salt at low pH
and the complexation of copper in the presence of excess oxalate
and at higher pH values. These calculations show that one must
very carefully control the experimental conditions in order to
obtain a 1:2:3 solid from a 1:2:3 Y:Ba:Cu solution. The
equilibrium data in the literature vary enough that an
experimental approach seems more profitable than extensive
calculations. In addition, there is no certainty that the metals
will behave independently.

Experimental

In the first part of the investigation, series of 1:2:3 metal
nitrate solutions were treated with quantities of oxalic acid
ranging from deficient to excess. The pH of sets of these
solutions was adjusted to various values by sodium or potassium
hydroxide. The solutions were filtered and analyzed for metal
content by flame emission. Some of the solids were fired and
examined for superconductivity.

    Other methods were investigated in attempts to obtain 1:2:3
solids. One that has been particularly successful was to treat
1:2:3 solutions of the nitrates with about 75% excess oxalic acid

to precipitate the bulk of the metals. The suspension was
evaporated to dryness under vacuum on a Buchi rotary evaporator.
This solid was then carefully heat treated in a platinum dish.
There was a very obvious exothermic reaction between nitrate and
oxalate at a temperature of about 310°C. This was preceded by a
darkening at around 300°C. Thus far it has been a totally controlled reaction, but this heat treatment should be done with care
using limited quantities of material. The same procedure was
attempted using chloride solutions in order to avoid the nitrate-
oxalate reaction, but the products have thus far not proven to be
superconducting.

Other investigations of this system are underway. We are
investigating the use of organic solvents to lower the solubility
of the barium oxalate. Organic bases rather than inorganic bases
are being tried which will avoid metal contamination.

The heat treatment of the oxalate samples has been fairly
standard. The initial calcining of the samples to 950°C in air
was done in either porcelain, alumina, or platinum containers.
The samples were reground, and then sintered to 930 to 950° in
oxygen. The first several hundred degrees of cooling was done
over several hours by turning down the furnace control by intervals
rather than simply turning it off. The samples were generally
pelletized for convenience in measurement before the final heat
treatment.

The presence of superconductivity was demonstrated by the
inductance method using equipment in the FSU Physics Department.
Measurements were also made by resistivity.

Results and Discussion

The oxalate precipitates are uniformly pale blue solids but the
superficial color gives no hint as to the metal composition. They
fire to give dark green to black solids. Of the mass that is lost
in the initial oxalate decomposition, most is lost by 350°C.
However, the remaining gray solid still has the basic bulk of the
original material. At temperatures between 800-900°C, a solid

state reaction occurs which results in a lot of shrinkage. Green
material almost always results when the composition is not precisely
1:2:3. Porcelain and quartz are extensively attacked. Material is
often deposited on quartz even when it is not in direct contact. It
is not at all obvious what the volatile material could be.

The results from the systematic pH controlled studies are
shown in Table 1. If no base is added, the solutions are colorless
indicating essentially complete removal of copper. However, barium
remains predominantly in solution. As the pH is increased, the
solutions take on the deep blue color of the copper oxalate complex
particularly when excess oxalic acid is present. Under these
conditions, it is, of course, the copper that is deficient in the
solid. However, there are several conditions under which solid of
close to 1:2:3 composition form. This is shown by both analysis
and the superconductivity of the fired pellets. A typical sample
has a Tc of 93.9 K with a narrow transition range.

An alternative to the problem of controlling the precipitation
conditions to make a solid of the proper composition is to avoid the
filtration step altogether. In this approach, the bulk of the
material was precipitated as the oxalate from nitrate solutions by
the addition of excess oxalic acid. The remainder was brought down
as the slurry was taken to dryness under vacuum with only slight
warming. No composition change can occur, and with luck, one can
maintain the benefit of finely divided material. There is a finely
divided surface for the final material to crystallize upon as it
does so under constant agitation. Ideally, the nitrate ion should
be volatilized as nitric acid (and trapped) under the conditions
used. However, it is clear that the nitrate was not completely
removed. On heating the solid, an exothermic reaction occurred at
about 300° that heated the material to a dull red heat. The
reaction was more than air oxidation as indicated by the evolution
of brown fumes. This reaction has thus far shown no signs of being
violent, but caution should be exercised.

The final product had the best Tc and the highest bulk density
of any material that we have made or that are generally reported.

TABLE 1   Oxalate Precipitation of Metals

Percent of various metals left in a 1:2:3 Y:Ba:Cu solution after
addition of various quantities of oxalate at various pH values

pH 0.7

%oxalate	yttrium	barium	copper
80%	8.3%	79%	2.0%
90	8.3	78	1.6
100	7.7	84	0.85
110	8.3	69	0.50
120	9.0	68	0.44

pH 3.5

80	8.6	89	1.3
90	6.1	59	2.0
100	3.2	32	1.2
110	1.6	14	7.5
120	1.3	5.6	16.1

pH 6.5

80	13.7	69	1.8
90	8.0	60	2.3
100	3.8	26	4.3
110	3.0	7.2	10.0
120	2.4	3.8	23.6

Initial starting composition was 3.12 mmole Y, 6.25 mmole Ba, and
9.38 mmole Cu in approximately 110 ml solution. The percent oxalate
is for complete precipitation of the three metals as standard 1:1 or
2:3 oxalate compounds.

A critical temperature range of 95.3 to 97.3K is normal and the density of the pellets is 5.2 to 5.7 g/cm^3, often in the higher part of that range. Most workers get bulk densities in the range of about 4.7 which is 70 to 75% of the crystal density. The theoretical density is 6.3.

We have attempted to substitute chloride solutions for the nitrates. One of the products of the precipitation reaction should be HCl which should readily distill under vacuum and give no problem with nitrate reaction. Surprisingly, the materials thus far have exhibited absolutely no superconductivity even though the metal composition must be 1:2:3.

ACKNOWLEDGMENTS

We wish to acknowledge the Jessie Ball duPont Religious, Charitable and Education Fund for financial assistance.

REFERENCES

1.   H. H. Wang, et.al., Inorg. Chem., 26, 1474 (1987).

2.   R. R. McCarley, Paper, ACS Meeting, New Orleans, 1987

3.   J. G. Bednorz, K. A. Muller, Z Phys. B: Condens. Matter. 64, 198 (1986).

11

High-Temperature Oxide Superconductor Thick Films

YONHUA TZENG Electrical Engineering Department, Auburn University,
200 Brown Hall, Auburn, Alabama 36849

INTRODUCTION

The recent discovery of ternary metal oxide ceramic
superconducting materials (Tc$\approx$98K) has led to a sudden burst of
research activity all over the world (1-9). The demonstrtion of
$10^5 A/cm^2$ critical current density at 77K (7) the boost the of
transition temperature to 155K by adding fluorine atoms to the
superconductor structure (10) and some evidence of room
temperature superconductors further promoted the promise and
interest of these materials. Processing of these materials into
useful forms such as wires, rods, tubes, films, ribbons, ... etc.
will make it possible for applications to high-field magnets,
power transmission lines, magnetic coil, microelectronics ...
etc. Among these special forms of superconductor, thick film is
probably one of the most promising approaches to the high
frequency, broad band electronic and magnetic shielding
applications.

The term "thick film" (11) has gained acceptance as the preferred generic description for microelectronics in which specially formulated pastes are applied and fired onto a ceramic substrate in a definite pattern and sequence to produce a set of individual components, such as resistors and capacitors, or a complete functional circuit. The pastes are usually applied using a silk-screen method. The high temperature firing matures the thick film elements and bonds them integrally to the ceramic substrate. Typically, the thickness of a thick film element will be 0.5 to 1 mil or more. This distinguishes it from thin film technology, where conductor thickness are generally much thinner. When active devices such as high frequency diodes and transistors are attached to a thick film network, the resulting product is known as a thick film hybrid circuit.

The advantages of thick film technology include higher performance, greater flexibility, outstanding reliability and cost-effectiveness. Resistors and capacitors in any of a wide variety of combinations, values, and characteristics can be constructed with the basic thick film materials and substrates. Thick film can even be employed in high power, high voltage and high frequency applications. Complete functional circuits from simple amplifiers to complex arrays containing many chip-type monolithic integrated circuits are possible in thick film form. Thick film circuits operating in the gigahertz range are routinely achievable.

Thick film hybrid microelectronic technology will be explored based on the ternary metal oxide ceramics $YBa_2Cu_3O_{6+x}$. The application of this new class of very exciting superconducting materials to microelectronics relies on the search for solutions to the problems in relatively low superconductor critical current density for non-single-crystalline materials and high metal-superconductor or semiconductor-superconductor contact resistance. Scientists in IBM have indirectly demonstrated the critical current density above 10^5 A/cm^2 in single crystalline YBa$_2$ Cu$_3$O$_{6+x}$ thin films which were electron beam evaporated and recrystallized on single crystalline strontium titanate substrates. In order to use this high current density capability, new and difficult techniques need to be invented so that single crystalline superconductor films can be deposited on other more interesting and lattice mismatched substrates, e.g., Si and GaAs.

One of immediate applications of polycrystalline superconducting materials is to form relatively dense thick films so that the critical current density is not a limiting factor. The main objective of this work is therefore devoted to the development of processing technology for hybrid system fabrication.

EXPERIMENTAL

Conventional screen printing techniques as well as direct painting on ceramic or other suitable substrates are investigated. This involves the choice of printing vehicles, the preparation of printing pastes, film thickness optimization, heat treatment of the printed films, and the formation of metal-superconductor and semiconductor-superconductor contacts.

Superconducting ternary metal oxide ceramics similar to that discovered by Professor M. K. Wu at University of Alabama in Huntsville, is prepared and tested. The YBa$_2$Cu$_3$O$_{6+x}$ ceramic is then ground into powder of about one micrometer in size and mixed with commercially available or custom-designed printing vehicles and/or thinners for pattern formation by means of a screen printer or direct painting. The $YBa_2Cu_3O_{6+x}$ thick film is sintered at $950^{\circ}C$ in an oxygen flowing tube for 12 hours and then removed from the furnace slowly.

RESULTS AND DISCUSSION

Shown in Figure 1 are an $YBa_2Cu_3O_{6+x}$ superconductor pellet, a directly painted superconductor thick film and screen printed superconductor patterns. These films are thicker than 1 mil. Four contacts are formed using silver paint at four corners of the samples in order to measure the resistance as a function of temperature. Shown in Figure 2 is a R v.s. T curve for the thick films. These superconductor thick films have superconductivity transition temperatures greater than $90^{\circ}K$ and the zero resistance state occurs at temperatures greater than 77K. XRD spectrum shows that the printed thick films have the same crystal structure as the superconductor pellets. This is shown in Figure 3. Some minor peaks due to the existence of multiple-phase structures can be seen in Figure 3.

The successful demonstration of high temperature superconductor thick films by means of a conventional printing technique can provide immediate applications of this material in desired patterns to hybrid microelectronics and microsensors. Further study is being done to investigate and optimize the printing and sintering processes in order to achieve a reliable superconductor-substrate interface and a higher superconductor critical current density.

Figure 1. Superconductor Samples. (a) Superconductor pellets,
           (b) Superconductor painted on an alumina substrate,
           (c) Superconductor screen printed on an alumina
           substrate.

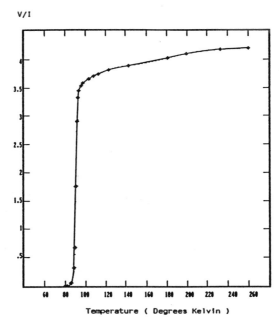

Figure 2. Four-point resistance measurement
           of a HTSC thick film.

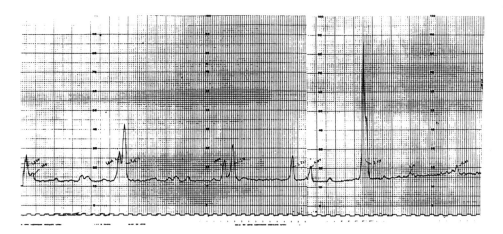

Figure 3. XRD spectrum of a HTSC thick film.

CONCLUSIONS

High temperature superconductor thick films based on the ternary
metal oxide $YBa_2Cu_3O_{6+x}$ have been successfully fabricated.   These
films have Tc greater than 90^OK and the zero resistance state
starts at temperatures greater than 77^OK.   High temperature
superconductor thick films of arbitrary shapes can be formed using
this technique.

ACKNOWLEDGEMENTS

This work was supported by the Alabama Microelectronics Science
and Technology Center. Professor M.   K.   Wu provided valuable
technical assistance.

REFERENCES

1. M.K. Wu, J.R. Ashburn, C.J. Torng, P.H. Hor, R.L. Meng, L. Gao, Z.J. Huang, Y.O. Wang and C.W. chu, Phys. Rev. Lett., 58, 908 (1987).

2. C. W. Chu, P.H. Hor, R.L. Meng, L. Gao, Z.J. Huang and Y.Q. Wang, Phys. Rev. Lett. 58, 405 (1987).

3. J.G. Bednorz and K.A. Muller, Z Phys. B 64, 189 (1986).

4. J.G. Bednorz, M. Takashige and K.A. Muller, Europhys. Lett., 3, 379 (1987).

5. Xhang Yuling, Xie Sishen, Cheng Xiangrong, Yang Qianshong, Liu guirong and Ni Yongming, J. Phys D. Appl. Phys. 20, 14, (1987).

6. S.K. Makik, A.M. Umarji, D.T. adroja, C.V. Tomy, Ram Prasad, M.C. Soni, Ashok Mohan, and C.K. Gupta, J. Phys. C: Solid State Phys. 20, L347. (1987).

7. P. Chandhari, R.H. Koch, R.B. Laibowitz, T.R. McGuire, and R.J. Gambino, Phys. Rev. Lett. (1987).

8. S.K. Dhar, P.L. Paulose, A.K. Grover, E.V. Sampathkumaran and V. Nagarajan, J. Phys. F: Met. Phys. 17, L105, (1987).

9. P.L. Paulose, V. Nagarajan, A.K. Grover, S.K. Dhar and E.V. Sampathkumaran, J. Phys. F: Met. Phys. 17, L91, (1987).

10. S.R. Ovshinsky, R.T. Young, D. D. Allred, G. DeMaggio, and G.A. Van der Leeden, Phys. Rev. Lett., 58, 2579, (1987).

11. C.A. Harper, ed. Handbook of Thick Film Hybrid Microelectronics McGraw-Hill, (1974).

12

Iodometric Titrations and Xanes: Two Perspectives of the Copper Valence in High-T$_c$ Superconducting Oxides

L. Soderholm, E.E. Alp, M.A. Beno, L.R. Morss, G. Shenoy, G.L. Goodman
Argonne National Laboratory, Argonne, IL 60439

INTRODUCTION

The recent observation of superconductivity above 30K in oxides containing copper [1,2] has opened up a new area of research. Central to the research direction is the question "What is special about these compounds which allows them to be superconducting at such high temperatures?" It appears that there are two main avenues which are being widely pursued in an effort to answer this question, the first based on structural arguments and the second on electronic considerations. The structural arguments focus on the necessity of Cu-O planes for 30K superconductivity, which are found in both La$_{2-x}$Sr$_x$CuO$_4$ (0≤x0.2) and YBa$_2$Cu$_3$O$_x$ (6.5≤x≤7.0), and the possible requirement of Cu-O chains for 90K superconductivity [3,4,5,6]. The electronic considerations focus on the necessity for "mixed valent" copper [3], ie. the presence of Cu^{2+}, Cu^{3+} and possibly even Cu^{1+} [7,8] or Cu^{4+} [9] before superconductivity can be achieved at high temperatures.

Key to understanding the importance of the "mixed valent" behavior of copper in these systems is the definition of the copper valence corresponding to a particular oxidation state. Since different techniques measure different things, it is in fact necessary to discuss the oxidation state of copper from more than one perspective. This can be done by distinguishing between the formal valence or oxidation state (e.g. $YBa_2Cu_2^{2+}Cu^{3+}O_7$), the nominal or average oxidation state (e.g. $Cu^{2.33+}$ in the above example), and the charge distribution measured experimentally, which may be different from the either of charge states described by the first two approaches. The concepts of nominal oxidation state and formal valences are important, providing a simple method of indicating how many electrons are available for bonding/band structure arguments. The actual oxidation state and valence, coming from the interpretation of experimental or calculated charge distributions, provides an avenue for discussing covalency, bonding, energy level diagrams and band structure calculations.

In this paper we compare and contrast results of the determination of copper charge states by chemical titration, which usually provides information about the nominal valence, and X-Ray Absorption Near-Edge Structure (XANES), which should provide evidence of the actual charge distribution in the solid.

EXPERIMENTAL TECHNIQUES
1. Chemical Titrations
The quantitative determination of copper has been well studied [10], since copper assays have important industrial and economic applications. A classical method of assay consists of dissolving the sample to produce an acidic solution containing Cu^{2+}. Potassium iodide (KI) is then added to reduce the divalent copper, and to produce iodine

$$Cu^{2+} + 2I^- \Rightarrow Cu^{1+}I \downarrow + 1/2I_2 \qquad (1)$$

The iodine can then be accurately titrated to determine the total copper in the dissolved sample. For superconducting oxides, a modification to this procedure has been proposed [11], in which the result of the above titration (for total copper) is compared to a similar experiment, in which KI is added to the solution prior to dissolving the copper. This permits the additional reaction

$$Cu^{3+} + 3I^- \Rightarrow Cu^{1+}I \downarrow + I_2 \qquad (2)$$

A combination of these two titrations permits the determination of the average or nominal Cu valence for values of Cu^{2+x} where $x \geq 0$. A further refinement of the basic method has been developed [12], in which the Cu^{2+} is complexed so that Cu^{3+} is the only oxidizing agent in solution, permitting a more accurate determination of small quantities of Cu^{3+}. Finally, dissolution of the sample in the presence of an oxidizing agent (e.g. Br_2) instead of a reducing agent also permits the determination of nominal valences of less than two [13].

2. X-ray Absorption

A copper absorption spectrum in the X-ray energy region has characteristic sudden increases in the absorption coefficient which result from the promotion of core electrons into excited (bound or unbound) states. For example, an isolated copper atom absorbs at 8979 eV X-rays as the result of the promotion of a 1s electron into a 3d state. This particular absorption is the copper K-edge. The exact location of this edge is influenced by the environment of the copper in the sample of interest. Within the solid state, crystal fields, covalent bonding, band formation and oxidation state or the physical manifestation of these effects, such as coordination number and geometry, can slightly alter the position, intensity and detailed features of the absorption spectrum.

X-ray Absorption Near Edge Structure (XANES) serves as a direct, single-ion probe of copper in the solid state, and is therefore a useful technique for looking at the charge density on copper in these high-T_c superconductors. It should be noted that this is a fast technique, with a time scale of approximately 10^{-16} seconds, and as a result it will be able to distinguish multiple copper oxidation states within a sample which are static or dynamic behavior.

RESULTS AND DISCUSSION
La_2CuO_4 and $La_{1.85}Sr_{0.15}CuO_4$

Titrations of La_2CuO_4 show that the bulk of the copper is divalent, but there is also a slight amount of Cu^{3+}, as listed in Table

Table 1. The results of chemical titrations of copper in solution.
The numbers in paranetheses are the standard deviations. These
errors represent only the error in titration, and do not include
possible errors from sample impurities.

Formula	Oxygen Coefficient y	Mean Cu Oxid. State	Cu^{1+}	Cu^{2+}	Cu^{3+}
		Chemical Methods			
			Percent		
La_2CuO_y	4.007(3)	2.014(6)	–	98.6	1.4
$La_{1.85}Sr_{0.15}CuO_y$	3.995(3)	2.140(6)	–	86.0	14.0
$YBa_2Cu_3O_y$	6.933(3)	2.289(3)	–	71.1	28.9
$YBa_2Cu_3O_y$	6.602(3)	2.068(3)	–	93.2	6.8
$YBa_2Cu_3O_y$	6.390(3)	1.926(3)	7.4	92.6	–

I. The detection of Cu^{3+} by chemical titration implies the presence of
defects in the form of a non-stoichiometry in the sample. Either the
presence of small amounts of La_2CuO_{4+x} or conversely some
$La_{2-x}CuO_4$ would explain these results. There is some evidence of
peroxide formation in this compound [14], which could lead to a
formulation La_2CuO_{4+x}. However, any additional oxygen forming a
peroxide bond is expected not to oxidize the copper, since the formal
charge of the peroxide ion is the same as the oxide anion. On the other
hand, there is evidence that $La_{2-x}CuO_4$ is a stable phase to at least
x=0.15 [15]. Furthermore this defect phase has been shown to be
superconducting, thus supporting the argument that some Cu^{3+} is
necessary for superconductivity. Work on the isostructural compound
$La_{2-x}CoO_4$ in which Co substitutes for Cu has shown that the
following reaction occurs upon prolonged heating under O_2 [16]:

$$La_{2-x}CoO_4 \Rightarrow LaCoO_3 + La_2O_3 \qquad (3)$$

There is no direct diffraction evidence that this type of reaction has
occurred in the copper compound, and $LaCuO_3$ has not been made at
ambient oxygen pressure. However it has been shown that the

prolonged heating of La_2CuO_4 under O_2 pressure results in a superconducting compound with a T_c of about 30K [17]. While this heat treatment could imply the formation of La_2CuO_{4+x}, the work on the isostructural cobalt compound suggests that a loss of some La_2O_3, though a phase separation of La_2CuO_4 analogous to (3) may explain this behavior. In this case it would be vacancies, instead of an alkaline earth, which substitute for La^{3+} and oxidize the copper.

The results of chemical titrations indicate that the replacement of La^{3+} by Sr^{2+} at concentrations up to $La_{1.85}Sr_{0.15}CuO_4$ produces Cu^{3+} which appears to scale with the Sr^{2+} concentration [12]. This observation is consistent with the replacement of La^{3+} by Sr^{2+} not causing an increase in oxygen vacancies which would produce a stoichiometry $La_{2-x}Sr_xCuO_{4-x/2}$. These titration results confirm other work which showed no evidence for the formation of oxygen defects at low Sr concentrations [14,18,19].

The XANES data reveal a much different story about the copper valence in La_2CuO_4, and the effect of Sr-doping. XANES measurements, described in detail elsewhere [20], were done on the copper K-edge. Experimental considerations make accurately measuring the absolute energy of the edge difficult, therefore a copper foil was run simultaneous to all experiments. The leading edge of the copper foil absorption serves as an internal energy standard which permits accurate comparisons of edges obtained in different measurements. These edges are then compared to the series of well characterized standards Cu_2O, CuO, and $KCuO_2$, in which the copper atoms assume a formally mono-, di-, or trivalent state respectively. Normalized spectra of the standards, together with that of La_2CuO_4 are presented in Figure 1. The general observation from the spectra of the standard compounds is that the absorption edge moves toward higher energy with increasing oxidation state of copper, a shift of about 2-3 eV/(valence change). The spectrum of La_2CuO_4 has two main features. The first is a strong peak located at roughly the position of the analogous peak in the CuO spectrum, and is attributed to the presence of divalent copper in the sample. The second main

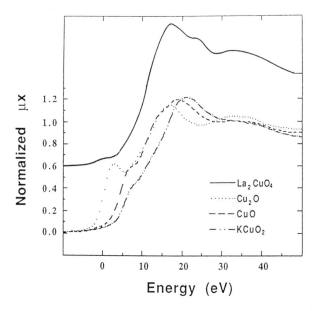

Figure 1. The Cu K-edge XANES of the formally mono- (Cu_2O), di-(CuO), and trivalent ($KCuO_2$) standards and La_2CuO_4.

feature, located about 3eV higher in energy, corresponds to the peak observed for $KCuO_2$, the standard for trivalent copper. Thus, the measured spectrum suggests that a substantial amount of <u>both</u> Cu^{2+} and Cu^{3+} states are already present in La_2CuO_4. A comparison of the XANES measurements on La_2CuO_4 with those of the doped superconductor $La_{1.85}Sr_{0.15}CuO_4$ is displayed in Figure 2. There is only a very minor difference between the two spectra (even less than expected for a 7.5% change in the average copper valence) which is a slight enhancement of the Cu^{3+} peak for the Sr-doped sample. These results indicate that the electronic structure, and therefore the measured charge state of Cu, is remarkably similar for both the doped and the undoped compound. Electronic structure calculations on molecular clusters [20] show that the electronic withdrawal caused by the replacement of La^{3+} by Sr^{2+} occurs primarily from the oxygen

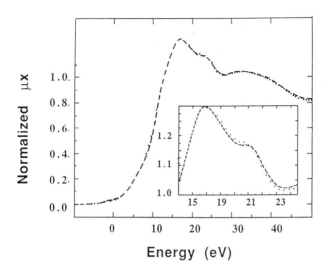

Figure 2. A comparison of the Cu K-edge XANES of La$_2$CuO$_4$ and La$_{1.85}$Sr$_{0.15}$CuO$_4$.

located axially to copper and closest to the Sr^{2+} site (O2 from ref 21). These calculations also indicated that there is only weak interactions between this axial oxygen and the planar Cu-O network, a result consistent with the long Cu-O bond.

Charge distributions obtained in the electronic structure calculations can be interpreted by charge density analysis [22,23]. In this way we are led to the conclusion that replacement of La^{3+} by Sr^{2+} serves more to oxidize its near neighbour oxygen ions than it does to oxidize Cu^{2+} to Cu^{3+} in this solid. The special solid state environment that stabilizes this unusual oxidation state of the oxygen ion disappears when the samples are dissolved. Then it is entirely possible that as the oxygen ion reverts to its formal valence of O^{2-}, the oxidation of Cu^{2+} to Cu^{3+} also is stabilized and subsequently measured by the iodide to iodine conversion. Hence it is not surprising that chemical titration and XANES give very different results.

YBa$_2$Cu$_3$O$_x$

While YBa$_2$Cu$_3$O$_x$ exhibits structural features and bulk properties
similar to those of La$_2$CuO$_4$, the XANES data indicate that the
electronic behavior of copper is different in these two systems. In
YBa$_2$Cu$_3$O$_x$, the nominal oxygen valence changes as a direct result of
changes in oxygen stoichiometry, thus avoiding the complication of
chemical doping. Chemical titrations, which determine Cu^{3+}
concentrations, now lead directly to the oxygen stoichiometry, so
Cu^{3+} scales with oxygen content. Normalized spectra of the copper
standard Cu$_2$O, together with the spectra of YBa$_2$Cu$_3$O$_{6.31}$,
YBa$_2$Cu$_3$O$_{6.46}$ and YBa$_2$Cu$_3$O$_{6.84}$ are shown in Figure 3, and discussed
in more detail elsewhere [24]. These results on the copper XANES
work follow much more closely with titration results on nominal
valence than did the XANES work on La$_{2-x}$Sr$_x$CuO$_4$. Unlike the
situation observed for La$_2$CuO$_4$, there is a notable change in the

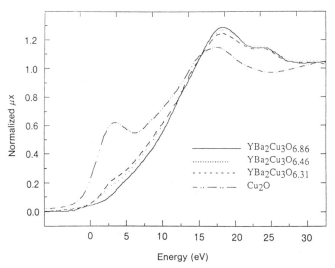

Figure 3. A comparison of the Cu K-edge XANES for YBa$_2$Cu$_3$O$_x$ as a
function of x. The edge can be seen to move to higher energy with
increasing x, consistent with an increase in the oxidation state of
copper.

details of the XANES data upon increasing nominal copper valence. The absorption edge of copper is seen to shift to higher energy as the oxygen content is increased. The edge of $YBa_2Cu_3O_{6.31}$ ($Cu^{1.87+}$) is at lower energy than is the edge for $YBa_2Cu_3O_{6.46}$ ($Cu^{1.99+}$) which is lower than $YBa_2Cu_3O_{6.84}$ ($Cu^{2.28+}$), as expected from the behavior of the standard compounds. Furthermore, the absorption coefficient in the x=6.31 compound shows a peak corresponding to Cu^{1+}, at approximately 3eV below the main edge, which coincides well with the spectrum of Cu_2O. Therefore, $YBa_2Cu_3O_{6.84}$ shows predominantly the presence of Cu^{2+} and Cu^{3+} while $YBa_2Cu_3O_{6.31}$ has Cu^{1+}, Cu^{2+} and Cu^{3+} charge states are present. It should be noted that $YBa_2Cu_3O_x$ has two crystallographically different copper atoms [6], and that XANES cannot directly distinguish between these sites. Therefore it is unclear if the three oxidation states observed for copper are in equilibrium, or if they are structurally isolated in the solid. It should also be noted that we have not observed any obvious features in the XANES data which we could directly attribute to the Cu^{4+}, although it has been proposed that Cu^{4+} occurs in this compound [9]. The lack of any known Cu^{4+} standard for comparison with our data prevents us from making a direct evaluation of our data.

It has been recently suggested that oxide superconductors, particularly the high-T_c materials (eg. $Ba(PbBi)O_3$, $La_{2-x}Sr_xCuO_4$ and $YBa_2Cu_3O_x$) all undergo a disproportionation reaction of the form:

$$M^{n+} \Rightarrow M^{(n-1)+} + M^{(n+1)+} \qquad (4).$$

For $Ba(PbBi)O_3$, this disproportionation would involve $Bi^{4+} \rightarrow Bi^{3+} + Bi^{5+}$ and for copper it would involve $Cu^{2+} \rightarrow Cu^{1+} + Cu^{3+}$. Whether there is a competition between superconductivity and disproportionation [25], or whether this reaction is directly involved in the mechanism for superconductivity [7] is unclear at this time. Our interpretation of the XANES results, particularly those for $La_{2-x}Sr_xCuO_4$, do not support this hypothesis. The absence of any significant amount of Cu^{1+} in either La_2CuO_4 or $La_{2-x}Sr_xCuO_4$, in either the normal or the superconducting state, rules out any substantial copper disproportionation in these materials. While reactions of the type shown in (4) are always observed for gold [26], and to some limited extent are observed for silver (the other elements in the 1B group), this reaction has never been demonstrated for copper. Furthermore, the observation of antiferromagnetism in La_2CuO_4 [27] indicates that local moments (unpaired spins) are formed on copper. This finding also suggests that a complete disproportionation of copper into Cu^{1+} and Cu^{3+} (both spin paired and diamagnetic) does not occur. The presence of Cu^{2+} and Cu^{3+}, as suggested by our XANES data, is consistent with the observed magnetic and chemical behaviour.

CONCLUSIONS

The valence of copper in the high-T_c superconductors $La_{2-x}Sr_xCuO_4$ and $YBa_2Cu_3O_x$ has been investigated by iodometric titrations and XANES. There is little apparent difference in the copper near edge spectra of La_2CuO_4 and the Sr-doped sample, although titration data show that the nominal oxidation state of copper does change with increasing dopant concentration. These apparently inconsistent findings are explained by calculations which show that it is the oxygen which is axial to the Cu-O planes which is most influenced by Sr-doping. On the other hand, the Cu K-edge is seen to shift to lower energy with decreasing oxygen content in $YBa_2Cu_3O_x$, consistent with the results of iodometric titrations.

ACKNOWLEDGMENTS

This work is supported by the U.S. Dept. of Energy, BES-Chemical and Material Sciences, under Contract #W-31-109-ENG-38.

REFERENCES

[1] J.G. Bednorz and K.A. Muller, Z. Phys. B64, 189 (1986).

[2] H. Takagi, S. Uchida, K. Kitazawa, S. Tanaka, Jpn. J. Appl. Phys. Lett. preprint.

[3] W.I.F. David, W.T.A. Harrison, J.M.F. Gunn, O. Moze, A.K. Soper, P. Day, J.D. Jorgensen, M.A. Beno, D.W. Capone II, D.G. Hinks, Ivan K. Schuller, L. Soderholm, C.U. Segre, K. Zhang and J.D. Grace, Nature 327, 310 (1987).

[4] D.C. Johnston, A.J. Jacobson, J.M. Newsam, J.T. Lewandowski, D.P. Gorhorn, D. Xie and W.B. Melon in Chemistry of High-Temperature Superconductors, ed. D.L. Nelson, M.S. Whittingham and T.F. George, ACS Symposium Series 351, American Chemical Society, p. 136 (1987).

[5] Ivan K. Schuller, D.G. Hinks, M.A. Beno, D.W. Capone II, L. Soderholm, J.-P. Locquet, Y. Bruynseraede, C.U. Segre and K. Zhang, Sol. St. Comm. 63, 385 (1987).

[6] M.A. Beno, L. Soderholm, D.W. Capone II, D.G. Hinks, J.D. Jorgensen, J.D. Grace, Ivan K. Schuller, C.U. Segre and K. Zhang, Appl. Phys. Lett. 51, 57 (1987).

[7] Arndt Simon, Angew. Chem. Int. Ed. Engl. 26, 579 (1987).

[8] Sleight in Chemistry of High-Temperature superconductors, ed. D.L. Nelson, M.S. Whittingham and T.F. George, ACS Symposium Series 351, American Chemical Society, p. 2, (1987).

[9] J. Yu, S. Massidda, A.J. Freeman and D.D. Koelling, Physics Lett. A122, 203 (1987).

[10] I.M. Kolthoff and R. Belcher, Volumetric Analysis Vol III, Interscience, N.Y., 1957, pp 347-365.

[11] D.C. Harris and Terrell A. Hewston, J. Solid State Chem. 69, 182 (1987).

[12] Evan H. Appelman, L.R. Morss, A.M. Kini, U. Geiser, A. Umezawa,
G.W. Crabtree and K.D. Carlson, Inorg Chem (in press).

[13] J.D. Jorgensen, M.A. Beno, D.G. Hinks, L. Soderholm, K.J. Volin, R.L.
Hitterman, J.D. Grace, Ivan K. Schuller, C.U. Segre, K. Zhang and
M.S. Kleefisch, Phys Rev B36(7), 3608 (1987).

[14] M.W. Shafer, T. Penny and B.L. Olson, Phys. Rev. B36(7), 4047
(1987).

[15] S.M. Fine, M. Greenblatt, S. Simizu and S.A. Friedberg, in
Chemistry of High-Temperature superconductors, ed. D.L. Nelson,
M.S. Whittingham and T.F. George, ACS Symposium Series 351,
American Chemical Society, p. 95 (1987).

[16] J.T. Lewandowski, R.A. Beyerlein, J.M. Longo and R.A. McCauley, J.
Am. Ceram. Soc., 69(9), 699 (1986).

[17] Jose Beille, R. Cabanel, C. Chaillout, B Chevallier, G. Demazeau,
F. Deslandes, J. Etourneau, P. Lejay, C. Michel, J. Provost, B.
Raveau, A. Sulpice, J.-L. Tholence and R. Tournier, Europhys.
Lett. 3(12), 1301 (1987).

[18] D.G. Hinks,in preparation.

[19] L. Soderholm, D.W. Capone II, D.G. Hinks, J.D. Jorgensen, Ivan K.
Schuller,J. Grace, K. Zhang and C.U. Segre, Inorg. Chim. Acta,
140(1-2) 167 (1987).

[20] E.E. Alp, G.K. Shenoy, D.G. Hinks, D.W. Capone, L. Soderholm, H.-B.
Schuttler, J. Guo, D.E. Ellis, P.A. Montano and M. Ramanathan,
Phys. Rev. B35, 7199 (1987).

[21] J.D. Jorgensen, H.-B. Schuttler, D.G. Hinks, D.W. Capone II, K.

Zhang, M.B. Brodsky and D.J. Scalapino, Phys. Rev. Lett. 58(10), 1024 (1987).

[22] R.S. Mulliken, J. Chem. Phys. 23, 1833 (1987).

[23] R.S. Mulliken, J. Chem. Phys. 23, 1841 (1987).

[24] E.E. Alp, L. Soderholm, G.K. Shenoy, D.G. Hinks, B.W. Veal and P.A. Montano, Physica, (submitted).

[25] A.W. Sleight, Chemtronics (in press).

[26] F.A. Cotton and G. Wilkinson, Advanced Inorganic Chemistry John Wiley and Sons, N.Y. pp. 1145 (1972).

[27] G. Shirane, Y. Endoh, R.J. Birgeneau, M.A. Kastner, Y. Hidaka, M. Oda, M. Suzuki and T. Murakami, Phys. Rev. Lett. 59(14), 1613 (1987).

13

Importance of the Interactions Between the Copper Atoms of the CuO$_2$ Layers Occurring via the Oxygen-Copper-Oxygen Atom Bridges of the CuO$_3$ Chains for the High-Temperature (T$_c$ > 90 K) Superconductivity of YBa$_2$Cu$_3$O$_{7-y}$

MYUNG-HWAN WHANGBO and MICHEL EVAIN Department of Chemistry, North Carolina State University, Raleigh, North Carolina 27695

MARK A. BENO and JACK M. WILLIAMS Chemistry and Materials Science Divisions, Argonne National Laboratory, Argonne, Illinois 60439

1. CRYSTAL STRUCTURE OF YBa$_2$Cu$_3$O$_{7-y}$

The high-temperature (T$_c$ > 90K) superconductor YBa$_2$Cu$_3$O$_{7-y}$,[1] as determined by powder neutron diffraction studies,[2] has an orthorhombic structure as shown in Figure 1. Substitution of Y^{3+} by lanthanide ions Ln^{3+} (e.g., Ln = Sm, Eu, Gd, Dy, Ho, Yb) hardly affects the high-temperature superconductivity,[3] so the structural unit essential for the high-temperature superconductivity is the two-dimensional (2D) Ba$_2$Cu$_3$O$_{7-y}$$^{3-}$ slab.[4] The ideal stoichiometry YBa$_2$Cu$_3$O$_7$ is possible when all the oxygen sites of Figure 1 are fully occupied. A stoichiometric Ba$_2$Cu$_3$O$_7$$^{3-}$ slab consists of two CuO$_2$ layers that sandwich one CuO$_3$ chain and two Ba atoms per unit cell. The copper atom coordination is 'square planar' for the CuO$_3$ chains, and is 'square pyramidal' for the CuO$_2$ layers. The O4 atoms of the CuO$_3$

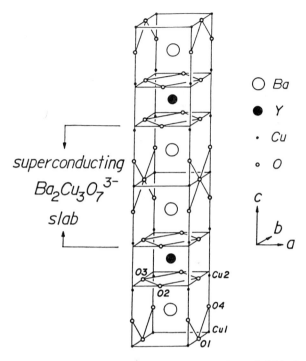

Figure 1.   Crystal structure of $YBa_2Cu_3O_{7-y}$ ($y \simeq 0.19$) determined
by powder neutron diffraction (Ref. 2a).

chains occupy the apical positions of the Cu2 atom 'square
pyramids', so the Cu2 atoms of one CuO_2 layer are linked to those
of the other CuO_2 layer via the O4–Cu1–O4 bridges of the CuO_3
chains.[4,5]

The oxygen atom vacancies of orthorhombic $YBa_2Cu_3O_{7-y}$
($y < 0.5$) occur primarily at the O1 positions of the CuO_3 chains,
i.e., the (0, 1/2, 0) sites of Figure 1.[2] Upon increasing
temperature, the orthorhombic phase gradually loses its oxygen
content of the Cu1 atom plane, and undergoes a phase transition
to a tetragonal structure as y increases beyond 0.5.[6] In
tetragonal $YBa_2Cu_3O_{7-y}$ ($y > 0.5$), both the (1/2, 0, 0) and

(0, 1/2, 0) sites are occupied by oxygen atoms with a site population less than 0.25, so the tetragonal phase no longer has the CuO$_3$ chains. With increasing temperature, the tetragonal phase gradually loses its oxygen content of the CuI atom plane eventually leading to the stoichiometry YBa$_2$Cu$_3$O$_6$,[6] which has no oxygen atoms in the CuI atom plane. The superconducting transition temperature T$_c$ of YBa$_2$Cu$_3$O$_{7-y}$ (y < 0.5) decreases with increasing y, and reaches zero beyond y > 0.5.[7] Therefore, the CuO$_3$ chains are crucial for the high-temperature superconductivity in YBa$_2$Cu$_3$O$_{7-y}$.

2. OXYGEN ATOM VACANCIES OF THE CuI ATOM PLANE AND THE Cu2-O4-
    CuI-O4-Cu2 LINKAGE

The oxygen atom environment of each Ba^{2+} ion is anisotropic in YBa$_2$Cu$_3$O$_{7-y}$ for all values of y = 0.0-1.0, because the CuI atom plane contains fewer O^{2-} ions than does the Cu2 atom plane.[8] The Coulomb attraction of the Ba^{2+} ion and the Coulomb repulsion of the O4 atom are smaller in magnitude with the O^{2-} ions of the CuI atom plane. Thus the CuI-O4 distance is shorter than the Cu2-O4 distance (e.g., 1.850(3) vs 2.303(3)A for y = 0.19), and the Ba...Ba distance is greater than the Ba...Y distance (e.g., 4.306(2) vs 3.688(2)A for y = 0.19).[2a,8]

In YBa$_2$Cu$_3$O$_{7-y}$ the extent of the anisotropy of the oxygen atom environment is enhanced with increasing y, because it is the oxygen atoms O1 that are primarily lost.[6,8] With increasing oxygen content, the Coulomb attraction of the Ba^{2+} ion and the Coulomb repulsion of the O4 atom toward the O^{2-} ions of the CuI atom plane become weaker, so that the capping oxygen atom O4 moves closer to, while the Ba atoms and hence the CuO_2 layers move farther away from, the Cu1 atom plane.[8] Consequently, as the oxygen content of the Cu1 atom plane decreases, the Cu1–O4 distance decreases while the Cu2–O4 distance increases as shown in Figure 2.[8b] Namely, the Cu2–O4–Cu1–O4–Cu2 linkages become longer with decreasing oxygen content, which will weaken the interlayer interaction in each $Ba_2Cu_3O_{7-y}^{3-}$ slab that occurs via the Cu2–O4–Cu1–O4–Cu linkages.

3. COPPER OXIDATION STATES

In stoichiometric $YBa_2Cu_3O_7$ there are two kinds of copper atoms, i.e., five-coordinate Cu2 and four-coordinate Cu1 atoms (1 and 2, respectively). In nonstoichiometric, orthorhombic $YBa_2Cu_3O_{7-y}$ ($0.5 > y > 0.0$), missing oxygen atoms O1 in the CuO_3 chains may lead to the three- and two-coordinate Cu1 atoms shown in 3 and 4, respectively. These Cu1 atoms are likely to occur in tetragonal $YBa_2Cu_3O_{7-y}$ ($y > 0.5$) as well. In principle, the four-coordinate Cu1 atom shown in 5 is possible for tetragonal $YBa_2Cu_3O_{7-y}$ due to random distribution (on a statistical basis of powder neutron diffraction measurements) of the oxygen atoms at the $(1/2, 0, 0)$ and $(0, 1/2, 0)$ sites. However, 5 is less likely to occur than is 2, since the Coulomb repulsion between the two oxygen atoms in the Cu1 atom plane will be greater for 5 due to the shorter O...O distance.[8b] For tetragonal $YBa_2Cu_3O_{7-y}$, the Cu1 atoms are unlikely to be either five- or six-coordinate since any right-angle arrangement of two oxygen atoms in the Cu1 atom

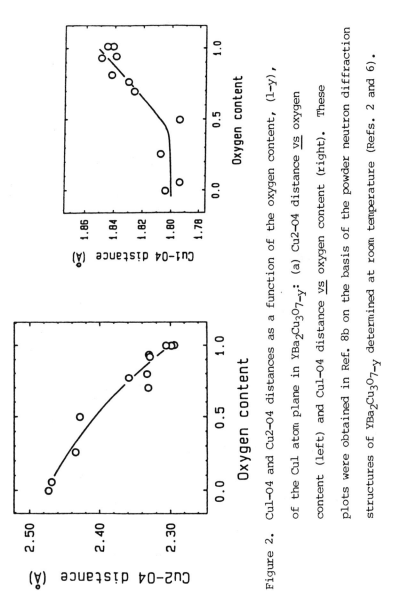

Figure 2. Cu1–O4 and Cu2–O4 distances as a function of the oxygen content, (1–y), of the Cu1 atom plane in YBa$_2$Cu$_3$O$_{7-y}$: (a) Cu2–O4 distance vs oxygen content (left) and Cu1–O4 distance vs oxygen content (right). These plots were obtained in Ref. 8b on the basis of the powder neutron diffraction structures of YBa$_2$Cu$_3$O$_{7-y}$ determined at room temperature (Refs. 2 and 6).

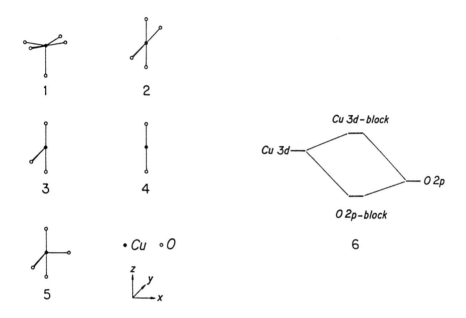

plane (e.g., **5**) will cause a strong Coulomb repulsion. Since several different coordinations are possible for the copper atoms, $YBa_2Cu_3O_{7-y}$ may contain copper atoms of different oxidation states. Since the term 'oxidation state' is often a source of controversy between chemists and physicists, it is necessary to state what we exactly mean by the oxidation state of the copper atoms in $YBa_2Cu_3O_{7-y}$.

Given neutral copper and oxygen atoms, the oxygen 2p (O 2p) level lies slightly lower, but the oxygen 2s (O 2s) level lies much lower, in energy than the copper 3d (Cu 3d) level. Thus, orbital interactions between the copper and oxygen atoms of $YBa_2Cu_3O_{7-y}$ occur primarily via the overlap between the Cu 3d and O 2p orbitals. As depicted in **6**, this orbital interaction gives rise to new energy levels. The O 2p-block level lies lower in energy than the O 2p level, since it has the Cu 3d and O 2p orbitals combined in-phase. The Cu 3d-block levels lies higher in energy than the Cu 3d level, because it has the Cu 3d and O 2p orbitals combined out-of-phase. Thus the Cu 3d- and O 2p-block levels have a mixed (i.e., 'hydridized') orbital character. In general, such d- and p-block levels of a transition metal compound are called, for short, the d- and p-levels of that compound, respectively.

For a transition metal compound, oxidation formalism is employed for the purpose of counting the number of electrons present in the d-block levels, which typically lie above the s- and p-block levels of the non-transion metal ligands. A convenient electron counting scheme for a copper oxide is to assume that all the oxygen s- and p-block levels are fully occupied, which is equivalent to assigning the oxidation state O^{2-} for oxygen, and that the remaining electrons occupy the Cu 3d-block levels. When there are 8, 9 and 10 electrons in the Cu 3d-block levels, the copper atoms are said to be in the oxidation states Cu^{3+}, Cu^{2+} and Cu^+, respectively, as if the electrons reside only on the copper atoms.

Figure 3 shows the Cu 3d-block levels calculated for the CuO_n (n=2-5) fragments 1-5 on the basis of their typical geometries found in orthorhombic $YBa_2Cu_3O_{7-y}$ (for 1 and 2) and tetragonal $YBa_2Cu_3O_{7-y}$ (for 3-5). Provided that the Cu^{2+} oxidation state can be assigned to 1 when all of 1-5 occur in

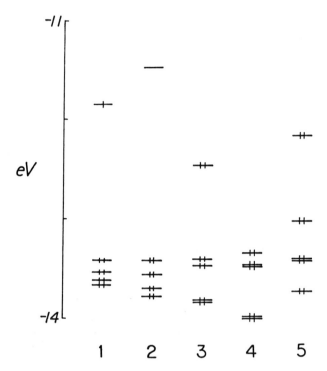

Figure 3. Cu 3d-block levels of the CuO_n (n=2-5) fragments (1-5)

calculated by the extended Huckel method.

$YBa_2Cu_3O_{7-y}$, the Cu^{3+} oxidation state should be assigned to 2

while the Cu^+ oxidation state should be assigned to 3-5. The

copper oxidation state is higher in 2 than in 1, although 2 has a

lower coordination number, because the Cu1-O4 distance is

considerably shorter than either the Cu2-O2 or the Cu2-O3

distance. Despite the fact that the Cu1-O4 distance is shorter

in 3-5 than in 2, the copper oxidation state is lower in 3-5.

This arises because the copper atoms of 3-5 can make 'nonbonding'

orbitals in two orthogonal directions (i.e., toward the missing

ligand positions) thereby lowering the resulting Cu 3d-block
levels.

4. BAND STRUCTURE OF A CuO$_2$ LAYER

Figure 4 shows the dispersion relation of the highest-
lying Cu 3d-block band of a flat CuO$_2$ layer, i.e., the x^2-y^2 band
if the coordinate x-, y-, and z-axes are taken along the a-, b-,
and c-axes of Figure 1 respectively. The orbital compositions of
this band at the wave vector points Γ, X and M are shown in 7, 8
and 9, respectively.[9] As the wave vector allows more O 2p
orbitals to combine with the Cu x^2-y^2 orbitals, the resulting
band levels are raised further in energy. Figure 5 shows the
density of states (DOS) calculated for the x^2-y^2 band and its
projections into the Cu x^2-y^2 and O 2p orbital contributions.
Clearly, the O 2p orbital contribution is small in the lower part
of the x^2-y^2 band, and both the O 2p and Cu x^2-y^2 orbital
contributions peak at the Fermi level. The ratio of the O 2p
contribution to the Cu x^2-y^2 contribution is about 1:2 at the
Fermi level, and increases beyond 1:2 above the Fermi level.

It is important to note that the x^2-y^2 band of a flat
CuO$_2$ layer is essentially identical to that of a flat CuO$_4$ layer[9]
found for another high-temperature superconductor La$_{2-x}$M$_x$CuO$_4$
(M = Sr, Ba; x $\simeq$ 0.1-0.2; T$_c$ $\simeq$ 30-40K).[10] This is so because the
axial oxygen atoms of CuO$_6$ octahedra that constitute the CuO$_4$
layer do not contribute to the x^2-y^2 band due to the δ-symmetry
of the x^2-y^2 orbital.

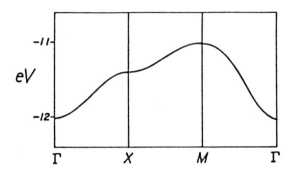

Figure 4.   Dispersion relation of the x^2-y^2 band of a flat CuO_2
            layer calculated by the extended Huckel tight-binding
            method (refs. 4, 5 and 9), where, in units of the
            reciprocal vectors $\underline{a}^*$, $\underline{b}^*$ and $\underline{c}^*$, the wave
            vectors Γ, X, Y and M are defined as: $\Gamma = (0,0,0)$,
            $X = (\underline{a}^*/2,0,0)$, $Y = (0, \underline{b}^*/2,0)$, and $M = (\underline{a}^*/2, \underline{b}^*/2,0)$.

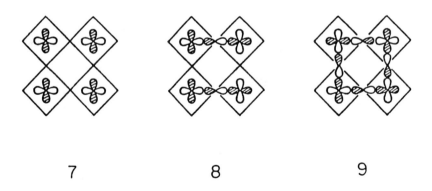

7   8   9

5. BAND STRUCTURE OF A $Ba_2Cu_3O_{7-y}^{3-}$ SLAB

When the coordinate x-, y- and z-axes are taken along the
$\underline{a}$-, $\underline{b}$- and $\underline{c}$-axes, respectively, the CuO_2 layers are parallel to
the xy-plane while the copper-oxygen planes of the CuO_3 chains

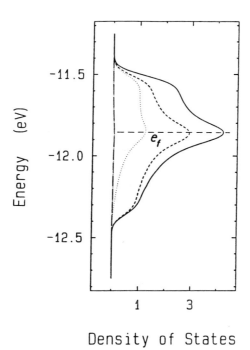

Density of States

Figure 5.   Density of states calculated for the x^2-y^2 band of

Figure 4. The total density of states is given by

the full line, and its projections into the Cu 3d, O 2p

and O 2s orbital contributions are given by the dash,

dot, and dash-dot lines, respectively.

are parallel to the yz-plane. Thus, in the CuO$_3$ chains, the z^2-y^2

orbital of Cu1 plays the role which the x^2-y^2 orbital of Cu2

plays in the CuO$_2$ layers.[4] Figure 6 shows the dispersion

relations of the top three d-block bands of a stoichiometric

Ba$_2$Cu$_3$O$_7$ layer calculated for its equilibrium structure,[4] where

the two nearly degenerate x^2-y^2 bands (represented by a thick

line) are solely derived from the CuO$_2$ layers, while the x^2-y^2

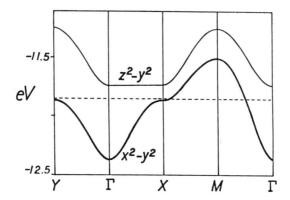

Figure 6.   Dispersion relations of the top three d-block bands of

a $Ba_2Cu_3O_7^{3-}$ slab calculated for its equilibrium

structure (Ref. 4).

band is solely represented by the CuO_3 chains. Since the z^2-y^2

orbital of CuI is raised in energy due to the short CuI-O4

distance, the z^2-y^2 band lies higher in energy than the x^2-y^2 bands.

There are two electrons to fill the three bands of Figure 6,

so that the two x^2-y^2 bands are half-filled, and the z^2-y^2 band

is empty. The bottom of the latter band lies close but slightly

above the Fermi level (0.11 eV). Figure 7 shows the DOS

calculated for the three bands of Figure 6 and its projections

into the Cu 3d and O 2p orbitals contributions of the CuO_2 layers

and the CuO_3 chain. It is evident from Figure 7 that the total

DOS profile is a superposition of the DOS for the one-dimensional

(1D) z^2-y^2 band and that for the 2D x^2-y^2 bands. The DOS value

at the Fermi level, $n(e_f)$, is calculated to be 8.85 electrons per

unit cell per eV.[4] Within a rigid band approximation, the Fermi

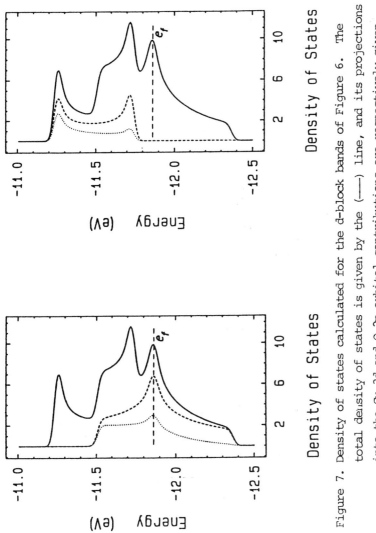

Figure 7. Density of states calculated for the d-block bands of Figure 6. The total density of states is given by the (———) line, and its projections into the Cu 3d and O 2p orbital contributions are respectively given by the (– – –) and (···) lines for the CuO$_2$ layers (left) and for the CuO$_2$ chain (right).

level of nonstoichiometric $YBa_2Cu_3O_{7-y}$ can be estimated by
putting 2+2y electrons (instead of 2) into the d-block bands of
Figure 6. This leads to a new Fermi level slightly higher than
the one shown in Figure 6, but still 0.06 eV below the bottom of
the x^2-y^2 band, and to a smaller $n(e_f)$ value (7.80 electrons per
unit cell per eV).[4]

The z^2-y^2 band of the CuO_3 chain is empty, and the half-
filled x^2-y^2 bands of the CuO_2 layers do not have any orbital
contributions from the CuO_3 chains. Within each superconducting
$Ba_2Cu_3O_{7-y}^{3-}$ slab, therefore, the Cu2 atoms of one CuO_2 layer do
not interact with those of the other CuO_2 layer although they are
linked by the Cu2-O4-Cu1-O4-Cu2 linkages. In essence, then, the
partially filled bands of a $Ba_2Cu_3O_{7-y}^{3-}$ slab calculated for its
equilibrium structure are identical in nature to the partially
filled band of a CuO_4 layer in $La_{2-x}M_xCuO_4$. This is apparently
puzzling in view of the long range order (i.e., the superconductivity)
that occurs at a high temperature ($T_c > 90K$), but arises from the
fact that, when the capping oxygen atoms O4 remain on the Cu2-Cu1
axis, as illustrated in **10**, all orbitals of the O4 atom have zero
overlap with x^2-y^2 orbital of the Cu2 atom due to the node of
this orbital along the Cu2-Cu1 axis.[4,5]

6. INTERLAYER INTERACTION VIA THE Cu2-O4-Cu1-O4-Cu2 LINKAGE

When the O4 atoms are displaced out of the Cu2-Cu1 axis
as illustrated in **11**, the z-orbital of the O4 atom, which
strongly overlaps with the z^2-y^2 orbital of Cu1, can have nonzero
overlap with x^2-y^2 orbital of Cu2.[5] Thus the Cu2 atoms of the

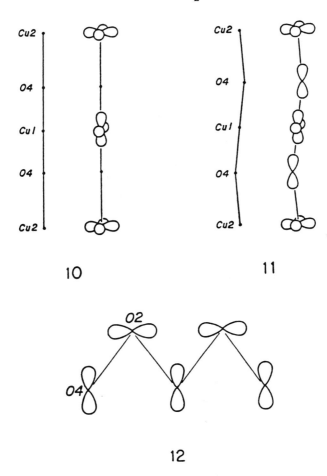

10   11

12

CuO$_2$ layers could interact electronically via the Cu2–04–Cu1–04–

Cu2 bridges. The band electronic structures of a Ba$_2$Cu$_3$O$_7$$^{3-}$ slab

calculated as a function of the 04 atom displacement along the c-

axis toward Cu2 (Δz) show the botton of the z^2-y^2 band of the

CuO$_3$ chain is lowered below the Fermi level when Δz > 0.04A.[5]

With Δz = 0.04A, the 04 atom displacement either along the a-axis

(Δx) or along the b-axis (Δy) by about 0.04A leads to appreciable

mixing between the z^2-y^2 and x^2-y^2 bands in the wave vector

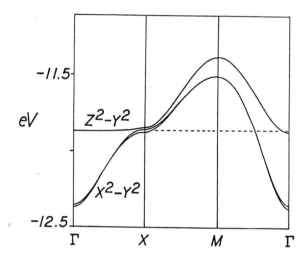

Figure 8. Dispersion relations of the top three d-block bands of a
           $Ba_2Cu_3O_7^{3-}$ slab calculated for the structure with the O4
           O4 atom displacements $\Delta z = \Delta x = 0.04A$ and $\Delta y = 0$ (Ref. 5).

region of X. Figure 8 shows the dispersion relations of the x^2-y^2
and z^2-y^2 bands of a $Ba_2Cu_3O_7^{3-}$ slab obtained for $\Delta z = \Delta x = 0.04A$
and $\Delta y = 0.$[5] Figure 9 shows the corresponding DOS calculated for
these bands and its projections into the Cu 3d and O 2p orbital
contributrions of the CuO_2 layers and the CuO_3 chain. The bottom
of the z^2-y^2 band lies near the middle of the x^2-y^2 bands where
the Fermi level occurs. Thus the $n(e_f)$ value, 11.87 electrons
per unit cell per eV, is considerably greater than the corresponding
value of 8.85 calculated for the case of $\Delta x = \Delta y = \Delta z = 0.$[5]

Figure 10 shows the Fermi surfaces calculated for the
three partially filled bands of Figure 8.[5] The Fermi surfaces of
the two x^2-y^2 bands are 2D, as expected, but so is the Fermi
surface of the z^2-y^2 band although it is primarily of the CuO_3

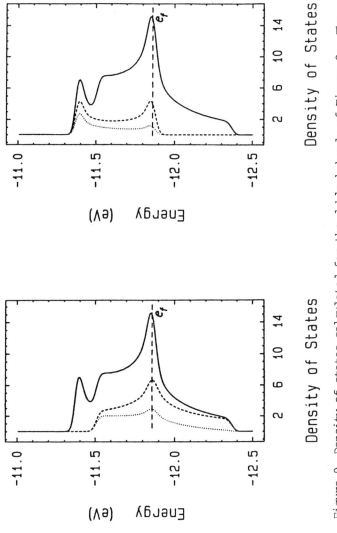

Figure 9. Density of states calculated for the d-block bands of Figure 8. The total density of states is given by the (——) line, and its projections into the Cu 3d and O 2p orbital contributions are respectively given by the (---) and (···) lines for the CuO$_2$ layers (left) and for the CuO$_2$ chain (right).

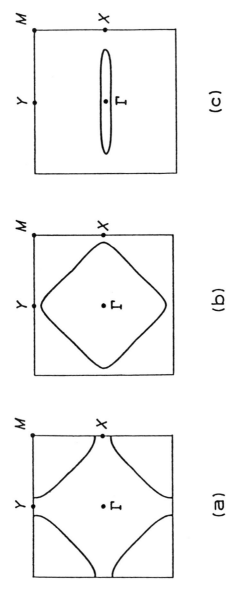

Figure 10. Fermi surfaces of the top three d-block bands of Figure 8: (a) lower x^2-y^2 band, (b) upper x^2-y^2 band, and (c) z^2-y^2 band (Ref. 5).

chains in character. This is caused by the fact that the z^2-y^2

band acquires a 2D character in the wave vector region of X by

mixing with the x^2-y^2 bands.[5] As indicated in **12**, the z-orbital

of O4 interacts not only with the x^2-y^2 orbital of Cu2 but also with

the x-orbital of O2 along the a-axis, which is perpendicular

to the chain direction. When the z^2-y^2 band is further lowered

by increasing the Δz value beyond 0.04A, the corresponding Fermi

level is raised more from the bottom of the z^2-y^2 band eventually

making this band filled for all the wave vectors along Γ to X.[5]

Then, the resulting Fermi surface of the z^2-y^2 band becomes 1D.

Depending upon the extent of the capping oxygen atom displacement,

therefore, the z^2-y^2 band can be either empty, partially filled

with a 2D Fermi surface, or partially filled with a 1D Fermi

surface. Partial filling of the z^2-y^2 band causes partial

emptying of the x^2-y^2 bands, so any O4 atom displacement that

lowers the z^2-y^2 band below the Fermi level slightly increases

the copper oxidation state in the CuO$_2$ layers but slightly

decreases that in the CuO$_3$ chains.[5]

Thus slight displacement of the O4 atom from its

equilibrium position gives rise to crucial changes in the band

electronic structure of a Ba$_2$Cu$_3$O$_{7-y}$$^{3-}$ slab. The root-mean-

square (rms) deviation of the O4 atom from its equilibrium

position is about 0.09A at room temperature,[2a] and similar values

are found at about 80K.[2b,c] Thus the O4 atom displacement of the

order of $\Delta x = \Delta y = \Delta z = 0.05$A should be easily accessible via the

lattice vibrational modes involving the O4 atoms. In the CuO$_3$

chains, the rms deviation of the O1 atom is highly anisotropic and is much greater than that of the O4 atom in the plane perpendicular to the Cu1-O1-Cu1 axis (e.g., 0.17, 0.06 and 0.16A, respectively, along the a-, b- and c- axes at room temperature).[2a] Band calculations reveal[5] that the O1 atom displacement of as large as 0.10A along a- or b-axis hardly lowers the z^2-y^2 band of the CuO_3 chain. This is so because the O1 atom is farther away from Cu1 than is O4 (i.e., Cu1-O1=1.943A $\underline{vs}$ Cu1-O4 = 1.850A), so the position of the z^2-y^2 band is insensitive to the O1 atom displacements.

For the long range order such as the high-temperature superconductivity to occur in each $Ba_2Cu_3O_{7-y}^{3-}$ slab, the copper atoms of the two CuO_2 layers must interact electronically. This can be achieved through the Cu-2-O4-Cu1-O4-Cu2 linkages with the help of the lattice vibrational modes involving the O4 atom displacement.[5]

7.   DEPENDENCE OF T_c ON OXYGEN CONTENT

If a sample of $YBa_2Cu_3O_{7-y}$ at a temperature T_Q is quenched (cooled rapidly), the resulting sample exhibits a T_c value which depends upon T_Q.[11] To a first approximation, the quenched sample would retain the oxygen content it had before quenching. On the basis of the known oxygen content $\underline{vs}$ temperature plot,[6a] therefore, one can deduce[8b] how the T_c value of $YBa_2Cu_3O_{7-y}$ varies as a function of its oxygen content from the available T_c $\underline{vs}$ T_Q data. As shown in Figure 11,[8b] the T_c value of $YBa_2Cu_3O_{7-y}$ is nearly constant at about 93K for $y \simeq 0.15$-0.25), decreases sharply with increasing y beyond 0.25 (the full line), and is zero for y > 0.5.

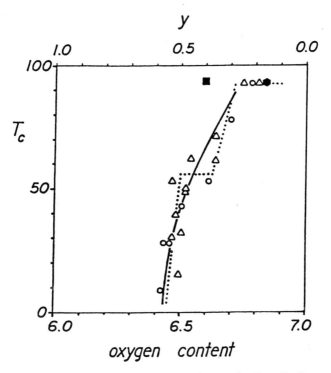

Figure 11. T_C versus overall oxygen content in YBa$_2$Cu$_3$O$_{7-y}$
(Ref. 8b), where the empty circles and empty
triangles represent the data of van den Berg et al.
(Ref. 11b) and Jorgensen et al. (Ref. 11a), respectively.
The filled square and filled hexagon refer to the data
of Katano et al. (Ref. 12) and Beno et al. (Ref. 2a),
respectively.

Apparently, the result of Katano et al.[12] deviates from
the above observation, which shows T_C = 94K at y = 0.40.   In
almost all YBa$_2$Cu$_3$O$_{7-y}$ samples the oxygen atom vacancies occur
primarily in the CuI atom plane, so that a change in the overall

oxygen content, 7-y, is nearly identical to that in the oxygen content of the Cul atom plane, 1-y. The $YBa_2Cu_3O_{6.6}$ sample of Katano et al. has substantial oxygen atom vacancies in the Cu2 atom planes, and its oxygen content of the Cul atom plane is 0.78,[12] which is only slightly smaller than that found in the $YBa_2Cu_3O_{6.81}$ sample by Beno et al.[2a] Therefore, what really controls the magnitude of T_C in $YBa_2Cu_3O_{7-y}$ is not the overall oxygen content but the oxygen content of the Cul atom plane.[8b] As already noted, the latter controls the length of the Cu2-O4-Cul-O4-Cu2 linkage and hence the magnitude of the interlayer interactions that occur via the Cu2-O4-Cul-O2-Cu4 linkages.[8b]

The data points of Figure 11 deviate from the full curve most significantly near T_C = 50-60K.[8b] Recent studies[7] on quenched $YBa_2Cu_3O_{7-y}$ samples reveal that the T_C vs oxygen content plot has two plateaus as schematically shown by the dotted lines in Figure 11 (e.g., $T_C \simeq$ 93K for $y \simeq$ 0.15-0.25; $T_C \simeq$ 55K for $y \simeq$ 0.40-0.50). It is important to understand how these two plateaus are related to the structural and electronic properties of $YBa_2Cu_3O_{7-y}$ that depend upon the oxygen content of the Cul atom plane. Suppose that a certain interlayer interaction which occurs via the Cu2-O4-Cul-O4-Cu2 linkage is essential for the high-temperature superconductivity (T_C > 90K). Then, weakening of this interlayer interaction, which may be brought about by lengthening the Cu2-O4-Cul-O4-Cu2 linkage, is expected to lower the T_C of $YBa_2Cu_3O_{7-y}$.[8b] If the Cu2-O4-Cul-O4-Cu2 linkage is lengthened beyond a certain value by removing enough Ol atoms from the Cu1 atom plane, the interaction between the CuO$_2$ layers

occuring via the Cu2–O4–Cu1–O4–Cu2 linkages may become so weak

that the CuO$_2$ layers of each Ba$_2$Cu$_3$O$_{7-y}^{3-}$ slab will be

independent of each other,[8b] as are the CuO$_4$ layers in La$_{2-x}$M$_x$O$_4$

whose T$_c$ value reaches 53K under an applied pressure of 12 kbar.[13]

Since the partially filled x^2-y^2 band of the CuO$_4$ layers

in La$_{2-x}$M$_x$CuO$_4$ is essentially identical to that of the CuO$_2$

layers of YBa$_2$Cu$_3$O$_{7-y}$, for a certain range of y values (e.g.,

y $\simeq$ 0.4–0.5), the CuO$_2$ layers of YBa$_2$Cu$_3$O$_{7-y}$ would be similar in

electronic structure to the CuO$_4$ layers of La$_{2-x}$M$_x$CuO$_4$ under an

applied pressure of 12 kbar. This would be responsible for the

lower plateau at T$_c$ $\simeq$ 55K in the T$_c$ $\underline{vs}$ oxygen content plot of

Figure 11.[8b]

8. CONCLUDING REMARKS

     From both the crystal and electronic structures of

YBa$_2$Cu$_3$O$_{7-y}$, it is evident that the interactions between the CuO$_2$

layers in each Ba$_2$Cu$_3$O$_{7-y}$ slab, which occurs via the Cu2–O4–Cu1–

O4–Cu2 linkages, are essential for the occurrence of the high-

temperature superconductivity (T$_c$ > 90K). When those linkages

are lengthened beyond a certain value by decreasing the oxygen

content of the Cu1 atom plane, the extent of the interlayer

interaction is reduced thereby making the CuO$_2$ layers decoupled

from each other and act independently just like the CuO$_4$ layers

of La$_{2-x}$M$_x$CuO$_4$. This accounts for the occurrence of the lower

plateau (T$_c$ $\simeq$ 55K, y $\simeq$ 0.40–0.50) in the T$_c$ $\underline{vs}$ oxygen content

plot. An interesting implication of these observations is that
the high-temperature superconductivity above 90K might originate
from the formation of Cooper pairs between the electrons of the
Cu2 atoms across the Cu2-O4-Cul-O4-Cu2 linkages as illustrated in
13. When such a Cooper pair formation is forbidden by
lengthening the Cu2-O4-Cul-O4-Cu2 linkages, Cooper pairs may be
formed between the electrons of the Cu2 atoms within each CuO_2
layer. The latter situation, identical in nature to that in the
CuO_4 layers of $La_{2-x}M_xCuO_4$, would lead to a lower T_c value (e.g.,
$T_c \simeq 55K$). Future theories aimed at explaining the high-
temperature superconductivity in $YBa_2Cu_3O_{7-y}$ must take into
account the importance of the interlayer interaction mediated by
the Cu2-O4-Cul-O4-Cu2 linkages.

    Recent self-consistent-field (SCF) band electronic structure
calculations[14] on the equilibrium structure of $YBa_2Cu_3O_7$ show
results essentially indentical to those discussed in the previous

secitons (i.e., those obtained by one-electron tight-binding band calculations[4,5]). However, those SCF calculations place the bottom of the CuO$_3$ chain z^2-y^2 band slightly above, not below, the Fermi level. This minor difference does not affect our conclusions concerning the importance of the O4 atom displacement for the electronic coupling between the CuO$_2$ layers via the Cu2-O4-Cu1-O4-Cu2 linkages. In SCF band structure calculations on YBa$_2$Cu$_3$O$_7$, if carried out, the z^2-y^2 band bottom will be raised above the Fermi level by shortening the Cu1-O4 bond since it will enhance the extent of antibonding between the Cu1 and O4 atoms in the z^2-y^2 level.

ACKNOWLEDGMENT

Work at North Carolina State University and Argonne National Laboratory were supported by the U.S. Department of Energy, Office of Basic Energy Sciences, Division of Materials Sciences under Grant DE-FG05-86-ER45259 and under Contract W31-109-ENG-38, respectively. We express our appreciation for computing time made available by DOE on the ER-Cray X-MP computer.

REFERENCES

1. Wu, M. K.; Ashburn, J. R.; Torng, C. J.; Hor, P. H.; Meng, R. L.; Gao, L.; Huang, Z. J.; Wang, Y. Q; Chu, C. W. Phys. Rev. Lett. 1987, 58, 908.

2. (a) Beno, M. A.; Soderholm, L.; Capone II, D.W.; Hinks, D. G.; Jorgensen, J. D.; Schuller, I.K.; Segre, C. U.; Zhang, K.; Grace, J. D. Appl. Phys.

Lett. 1987, 51, 57.

(b) Greedan, J. E.; O'Reilly, A.; Stager, C. V. Phys. Rev. B: Condens. Matter 1987, 37, 8770.

(c) Capponi, J. J.; Chaillout, C.; Hewat, A.W.; Lejay, P.; Marezio, M.; Nguyen, N.; Raveau, B.; Soubeyroux, J. L.; Tholence, J. L.; Tournier, R. Europhys. Lett., 1987, 3, 1301.

(d) David, W. I. F.; Harrison, W. T. A.; Gunn, J. M. F.; Moze, O.; Soper, A. K.; Day, P.; Jorgensen, J. D.; Beno, M. A.; Capone II, D. W.; Hinks, D.G.; Schuller, I. K.; Soderholm, L.; Segre, C. U.; Zhang, K.; Grace, J. D. Nature, 1987, 327, 310.

(e) Beech, F.; Miraglia, S.; Santoro, A.; Roth, R.S. Phys. Rev. Lett. (submitted).

(f) Hewat, A. W.; Capponi, J. J.; Chaillout, C.; Marezio, M.; Hewat, E. A. Nature (submitted).

3. (a) Fisk, Z.; Thompson, J. D.; Zirngiebl, E.; Smith, J.L.; Cheong, S.-W. Solid State Commun. 1987, 62, 743.

(b) Porter, L. C.; Thorn, R. L.; Geiser, U.; Umezawa, A.; Wang, H. H.; Kwok, W. K.; Kao, H.-C.; Monaghan, M. R.; Crabtree, G. W.; Carlson, K. D.; Williams, J. M. Inorg. Chem. 1987, 26, 1645.

(c) Engler, E. M.; Lee, V. Y.; Nazzal, A. I.; Beyers, R. B.; Lim, G.; Grant, P. M.; Parkin, S. S. P.; Ramirez, M. L.; Vazquez, J. E.; Savoy, R. J. J. Am. Chem. Soc. 1987, 109, 2848.

4. Whangbo, M.-H.; Evain, M.; Beno, M. A.; Williams, J. M.
    Inorg. Chem., **1987**, *26*, 1831.

5. Whangbo, M.-H.; Evain, M.; Beno, M. A.; Williams, J. M.
    Inorg. Chem., **1987**, *26*, 1832.

6. (a) Jorgensen, J. D.; Beno, M. A.; Hinks, D. G.; Soderholm, L.;
        Volin, K. J.; Hitterman, R. L.; Grace, J. D.; Schuller, I. K.;
        Segre, C. U.; Zhang, K.; Kleefisch, M. S. Phys. Rev. B:
        Condens. Matter (in press).

    (b) Miraglia, S.; Beech, F.; Santoro, A.; Tran Qui, D.;
        Sunshine, S. A.; Murphy, D. W. Mat. Res. Bull.
        (submitted).

    (c) Renault, A.; McIntyre, G. J.; Collin, G.; Pouget, J.-P.;
        Comes, R. J. Physique, **1987**, *48*, 1407.

    (d) Santoro, A.; Miraglia, S.; Beech, F.; Sunshine, S. A.;
        Murphy, D. W.; Schneemeyer, L. F.; Waszczak, J. V.
        Mat. Res. Bull. (submitted).

    (e) Onoda, M.; Shamoto, S. -I.; Sato, M.; Hosoya, S.
        Jpn, J. Appl. Phys., **1987**, *26*, L876.

7. (a) Veal, B. W.; Jorgensen, J. D.; Crabtree, G. W.;
        Kwok, W.; Umezawa, A.; Paulikas, A. P.; Morss,
        L. R.; Appelman, E. H.; Nowicki, L. J.; Nunez, L.;
        Claus, H. International Conference on Electronic
        Structure and Phase Stability in Advanced Ceramics;
        August 17-19, **1987**; Argonne National Laboratory,
        Argonne, Illinois, USA.

    (b) Johnston, D. C.; Jacobson, A. J.; Newsam, J. M.;

Lewandowski, J. T.; Goshorn, D. P.; Xie, D.; Yelon, W. B. Symposium on INorganic SUperconductors; American Chemical Society National Meeting; August 31-September 4, 1987; New Orleans, Louisiana, USA.

(c) Cava, R. J.; Batlogg, B.; Chen, C. H.; Rietman, E.A.; Zahurak, S. M.; Werder, D. Phys. Rev. (submitted).

8. (a) Evain, M.; Whangbo, M.-H.; Beno, M. A.; Geiser, U.; Williams, J. M. J. Am. Chem. Soc. (submitted).

(b) Whangbo, M.-H.; Evain, M.; Beno, M. A.; Geiser, U.; Williams, J. M. Inorg. Chem. (in press).

9. Whangbo, M.-H.; Evain, M.; Beno, M. A.; Williams, J. M. Inorg. Chem., 1987, 26, 1829.

10. (a) Bednorz, J. G.; Muller, K. A. Z. Phys. B.: Condens. Matter, 1986, 64, 189.

(b) Uchida, S.; Takagi, H.; Kitazawa, K.; Tanaka, S. Jpn. J. Appl. Phys. Part 2, 1987, 26, L1.

(c) Takagi, H.; Uchida, S.; Kitazawa, K.; Tanaka, S. Jpn. J. Appl. Phys. Part 2, 1987, 26, L123, L218.

(d) Cava, R. J.; van Dover, R. B.; Batlogg, D.; Rietman, E. A. Phys. Rev. Lett., 1987, 58, 408.

(e) Capone II, D. W.; Hincks, D. G.; Jorgensen, J. D.; Zhang, K. Appl. Phys. Lett., 1987, 50, 543.

11. (a) Jorgensen, J. D.; Veal, B. W.; Kwok, W. K.; Crabtree, G. W.; Umezawa, A.; Nowicki, L. J.; Paulikas, A. P. Phys. Rev. Lett. (submitted).

(b) van den Berg, J.; van der Beek, C. J.; Kes, P. H.;

Nieuwenhuys, G. J.; Mydosh, J. A.; Zandbergen, H. W.;
van Berkel, F. P. F.; Steens, R.; Ijdo, D. J. W.
Europhys. Lett. (submitted).

12. Katano, S.; Funahashi, S.; Hatano, T.; Matsushita, A.;
Nakamura, K.; Matsumoto, T.; Ogawa, K. Jpn. J. Appl.
Phys. Part 2, 1987, 26, L1046.

13. (a) Chu, C. W.; Hor, P. H.; Meng, R. L.; Gao, L.; Huang, Z. J.;
Wang, Y. Q. Phys. Rev. Lett, 1987, 58, 405.

    (b) Chu, C. W.; Hor, P. H.; Meng, R. L.; Gao, L.; Huang, Z. J.
Science (Washington, D.C.), 1987, 235, 567.

14. (a) Mattheis, L.F.; Hamann, D. R. Solid State Commun.
(in press).

    (b) Herman, F.; Kasowski, R. V.; Hsu, W. Y. Phys. Rev. B
(submitted).

14

Crystal Structure and Lattice Vibrations of the Ceramic Superconductor La$_{1.85}$Sr$_{0.15}$CuO$_4$: Neutron Scattering Studies

P. DAY   Oxford University, Inorganic Chemistry Laboratory, South
Parks Road, Oxford OX1 3QR, U.K.

High resolution powder neutron diffraction shows that the ceramic
superconductor La$_{1.85}$Sr$_{0.15}$CuO$_4$ (T$_c$ = 36K) is tetragonal
(I4/mmm, a = 3.7748(1); c = 13.2213(3)Å) at room temperature but
below 180K an orthorhombic distortion is found, which increases
smoothly down to 30K (Abcm, a = 5.3432(1), b = 5.3193(1), c =
13.1785(3)Å), decreasing again at 22K.   There is no detectable
anomaly in the evolution of the structure with temperature in the
vicinity of T$_c$.   Incoherent inelastic neutron scattering has
been used to measure the lattice vibrations in this compound and
the antiferromagnetic (T$_N$~250K) metal La$_2$CuO$_4$ from 0-400 cm^{-1} at
temperatures between 200K and 30K.   For the superconductor the
vibrational spectrum, which is temperature independent over the
whole range has major peaks at 96, 160, 248, 320, 440 cm^{-1} while
La$_2$CuO$_4$ has peaks at (65), 92, 144, 224, 304, 480 cm^{-1}.   Only the
increased scattering at 65 cm^{-1} in La$_2$CuO$_4$ has no analogue in the
superconductor.   Comments are made about possible mechanisms of
superconductivity in the quaternary Cu oxides.

1.   INTRODUCTION

The first of the new superconducting copper oxide ceramics to
be identified earlier this year were those whose structures are

based on the K_2NiF_4 lattice, namely $La_{2-x}M_x^{II}CuO_{4-y}$, with x about 0.15, M^{II} = Ca,Sr,Ba and y tending to zero in the best samples.[1] In recent months these compounds have tended to be overshadowed in interest by the oxygen-deficient perovskite series $M^{III}Ba_2Cu_3O_{7-\delta}$ (M^{III}=Y or 4f block element, $\delta \sim 0.1$) because the latter have superconducting onsets around 90K, compared with 35-40K in the K_2NiF_4 series.[2] Nevertheless, the 40K series retain their importance as prototype systems through which to try and understand the fundamental mechanism of the superconductivity for a number of reasons. Most important is the relative simplicity of their crystal structures, which contain only one Cu coordination site, compared with two in the 1:2:3 compounds. Also the synthesis of samples with y approaching zero means that the conduction band filling is determined exclusively by the La:M^{II} ratio, which can easily be changed chemically. In contrast, there has been much debate [3] about which oxygen sites gain or lose anions in the 1:2:3 lattice, and the influence this has on the physical properties. Finally the existence of a parent compound, La_2CuO_4, with its own unusual magnetic behaviour [4] signals the close (and potentially controllable) competition between magnetism and superconductivity. For all these reasons we have chosen to carry out detailed studies of the crystal structure and vibrational excitations of one of the 40K superconductors, $La_{1.85}Sr_{0.15}CuO_4$.

Of great interest is the way in which the crystal structure evolves with temperature, especially in the question of lattice distortions that might be related to the transition from metallic to superconducting. Since the ceramic superconductors are usually prepared by high temperature sintering they have only recently started to become available as single crystals. For this reason we studied the temperature dependence of the crystal structure of $La_{1.85}Sr_{0.15}CuO_4$ from 300 to 20K by high resolution powder neutron diffraction. Because of the small atomic number of oxygen relative to the other atoms present, X-ray diffraction is not sensitive enough to show the full detail of the crystal structures. Neutron scattering is therefore the method of choice to study both structure and excitations. The availability of high resolution powder neutron diffractometers and inelastic scattering spectrometers in Europe at ILL Grenoble and the ISIS pulsed source at the Rutherford Laboratory in the UK has enabled definitive structural data and phonon densities of states to be obtained over a wide range of temperature. It is the purpose of the present chapter to summarize these results and indicate what can be learned from them about the superconductivity mechanism. It is also pertinent to make some remarks about preparative methods, homogeneity and micro-structure of these materials.

2. PREPARATION AND CHARACTERIZATION

One of the disadvantages of neutron scattering is that relatively large quantities of sample are needed. In the present case, about 10g were used for the diffraction measurement and 20g for the inelastic neutron scattering. Clearly it is of vital importance that the sample should be completely pure and homogeneous: this means free of unreacted starting materials or other phases, and with a uniform distribution of La and Sr, both between and within crystallites. In $YBa_2Cu_3O_7$ the Y^{3+} and Ba^{2+} are completely segregated into layers ...YBaBaYBaBa... perpendicular to the c-axis (5). Is there any comparable segregation in $La_{2-x}M_xCuO_4$?

The customary preparative method for ternary oxide phases (the so-called 'ceramic' method) is to grind together the parent binary oxides or carbonates and fire them close to the melting point, relying on cation diffusion to form the thermodynamically favoured product at a convenient rate. Even after repeated regrinding and firing this method often results in small amounts of unreacted CuO together, in the early preparations,(6) with other perovskites such as $La_4BaCu_5O_{13+x}$.(7) For this reason we have used alternative preparative methods based on coprecipitation of carbonates or ceramic sol-gel techniques. Like others (8a), we have found that careful pH control is needed to ensure that the stoichiometry of the La,Sr,Cu carbonate precipitate matches that of the original solution. Buffering the solution of mixed metal nitrates, (initially pH4) to pH7 with an acetic acid/Na acetate buffer avoids bicarbonate formation and ensures complete Sr^{2+} precipitation.

Alternatively, a standard sol-gel precipitation-decomposition process (8b) avoids the necessity of filtering large quantities of precipitate. It starts from aqueous solutions of the metal nitrates, powdered citric acid and ethylene glycol. After a vigorous exothermic reaction, the resulting gel is further heated to decompose it to a black solid, which is then ground and formed into pellets under 10 tons pressure. The latter are annealed in air at 1050°C for 20 hours. X-ray powder diffraction of these samples shows no detectable trace of any other phase and all peaks index in the tetragonal I4/mmm space group with a=3.781(2); c=13.232(1)Å. Four probe a.c. conductivity of a block cut from such a sample of composition $La_{1.83}Sr_{0.17}CuO_4$ is shown in Fig. 1. The superconducting onset is at 42K with midpoint at 38K, 10-90% transition width of 2K with zero resistance at 33K. A.c. susceptibility measurements indicate a volume fraction of 70(10)% superconducting material at 4.2K by comparison with Pb. These characteristics compare favourably with those of samples prepared by oxide grinding (1), the narrower transition width indicating the superiority of 'wet' techniques in achieving homogeneous samples.

After overall stoichiometry and purity, further questions of chemical characterization concern homogeneity of composition (La,Sr segregation) and microstructure (presence and ordering of defects). Energy dispersive X-ray analysis (EDX) carried out in an electron microscope showed no detectable variation of La:Sr:Cu ratio over a large number of crystallites. High resolution

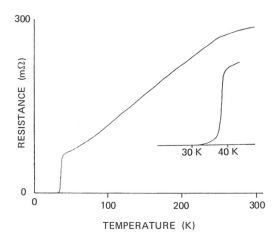

Fig. 1   Four-probe a.c. conductivity of La$_{1.83}$Sr$_{0.17}$CuO$_4$.

lattice images (resolution 0.14nm) reveal excellent crystallographic order over large areas of material, with no indications
either of point defects or intergrowths of other phases such as
has been observed (9a) in the La$_{n+1}$Cu$_n$O$_{3n+1}$ series of
compounds.   Unfortunately, of course, electron microscopy is
relatively insensitive to oxygen so the key question of the extent
and arrangement of the anion vacancies can only be addressed
through TEM by the effect they have on the heavy cation distribution.   On that evidence there appears to be no detectable ordering
of anion vacancies in the La$_{2-x}$Sr$_x$CuO$_{4-y}$ series.   As we
shall see below, neutron diffraction allows us to refine an
average site occupancy, but of course gives no view of the local
structure.

3.   CRYSTAL STRUCTURES

     At room temperature, the crystal structure of the parent
compound La$_2$CuO$_4$, first determined some years ago by X-ray

diffraction (9b) exhibits a small orthorhombic distortion (to Abcm) from the tetragonal (I4/mmm) K_2NiF_4 structure, though it becomes tetragonal above 554K.(7)   On doping with M^{2+} the tetragonal-to-orthorhombic transition temperature T_d falls, till in the superconducting regime (0.1$\leq$x$\leq$0.2) the lattice is tetragonal at room temperature. The basic feature of the K_2NiF_4 structure (Fig. 2) is the square planar layer of corner sharing NiF_6 octahedra. In La_2CuO_4 this layer is puckered along the $[110]$ direction of the I4/mmm so that the NiF_6 undergo an essentially rigid tilt, with the basal plane O(2) being shifted alternately above and below the mean plane of the Cu. With respect to the superconducting phases, the questions to be answered are whether this, or any other, distortion occurs below room temperature, and if so, whether $T_d(x)$ bears any relationship to the onset of superconductivity (T_c). Further questions concerning the order of the transition at T_d, the extent of occupancy of the O sites and any changes of bond lengths associated with the increase in the average Cu oxidation state.

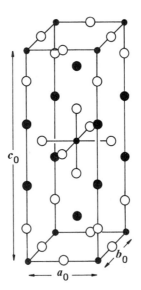

Fig. 2   The K_2NiF_4 structure: small filled circles Ni, large filled circles K, large empty circles F.

The use of powder neutron diffraction was forced on us by the unavailability of single crystals at the time when the experiments described here were performed (March 1987). Nevertheless, the technique is in fact a very convenient one for monitoring temperature dependent structure changes because, at least on the High Resolution Powder Diffractometer at the ISIS Pulsed Neutron Source, (SERC Rutherford Appleton Laboratory, Chilton, UK) the flux is such that diffraction profiles of excellent peak-to-background quality can be obtained in about 2 hours, as can be seen from Fig. 3.

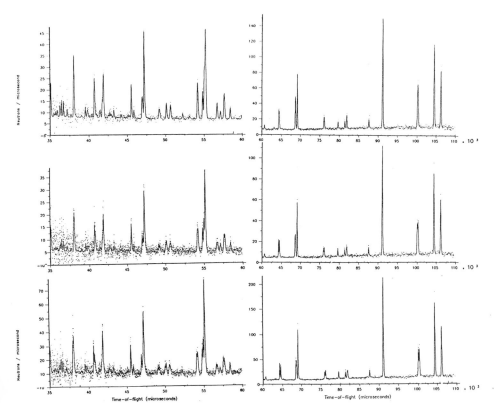

Fig. 3   Powder neutron diffraction profiles of La$_{1.85}$Sr$_{0.15}$CuO$_4$ at (from top to bottom) 190, 110 and 30K. The lines through the data points are the fitted profiles.

We measured thirteen such profiles of $La_{1.85}Sr_{0.15}CuO_4$ at temperatures from 300 to 22K, and fitted them by the Rietveldt profile refinement method (10) to give the unit cell constants shown in Fig. 4, and the atomic positional parameters and site occupancies summarized in Table 1. Although the latter data are available for all thirteen temperatures, we only list those for 22, 30 and 300K.

There was no evidence of any deviation from tetragonal symmetry down to 180K, and the structure refinements at higher temperatures were therefore carried out in the I4/mmm space group, with two formula units per unit cell. The CuO_6 octahedra are strongly elongated along the c-axis, in other words the Cu-O bond length within the basal plane (1.88995(7)Å at 300K) is considerably shorter than the axial Cu-O bond length (2.4043(3)Å at 300K) so much so that the apical oxide ion O(1) is actually closer to the La(Sr) (2.3697(3)Å at 300K) than to the Cu. Refinement of the room temperature data to include anisotropic temperature factors reveals anomalously large values of B_{11} (1.56(8)) for the apical oxygen O(1) and of B_{33} (1.3(2)) for the equatorial oxygen O(2). We think it significant that this is exactly what would be anticipated if a phonon associated with a rigid tilting of the CuO_6 units about the [110] direction, which becomes the b-axis of the enlarged orthorhombic cell at low temperature, were becoming 'soft'.

Figure 3 shows sample diffraction profiles measured at 190, 110 and 30K, on which the dots represent the experimental data points and the lines are fitted by the Rietveldt refinement. Close inspection shows that with decreasing temperature some of the peaks (e.g. those near 64, 76 and 100μs) become split, signalling a distortion to a lower symmetry. The effect is shown in greater detail in Fig. 5 for the (040, 400) (I4/mmm) peak from 300K to 22K. From 300 to 190K the lattice constants and bond lengths contract, but the unit cell remains tetragonal. However, below 180K the magnitude of the distortion increases smoothly with

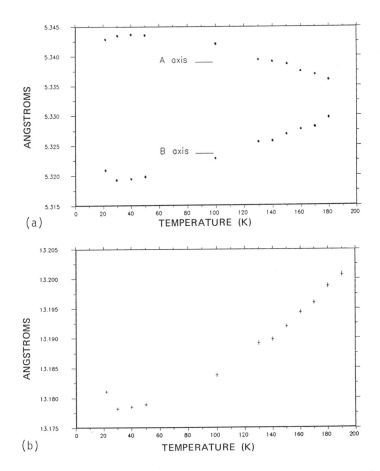

(a)

(b)

Fig.4   Temperature dependence of the unit cell parameters of
        La$_{1.85}$Sr$_{0.15}$CuO$_4$ (a)a and b, (b) c.

decreasing temperature down to 30K. Successful refinements of all
the data sets below 180K were carried out in the orthorhombic Abcm
space group, assuming that the distortion occurs by alternate
tilting of the CuO$_6$ units along the [110] direction of the parent
I4/mmm unit cell, as in La$_2$CuO$_4$. The maximum tilt angle reached
at 30K is 3.0(5)$^\circ$, which should be compared with 5.5° in La$_2$CuO$_4$.
In Fig. 4 is plotted the temperature dependence of the (Abcm) cell
parameters. Data on the magnitude of the orthorhombic distortion

Table 1. Atomic coordinates, temperature factors and O site occupancies for $La_{1.85}Sr_{0.15}CuO_4$ (22 & 30K Abcm; 300K I4/mmm).

		22K	30K	300K
La,Sr	x	0.0054(12)	0.0081(17)	0
	z	0.3614(2)	0.3614(2)	0.3611(2)
	B	0.16(12)	0.01(15)	0.56(7)
Cu	B	0.2(2)	0.1(1)	0.44(8)
O(1)	x	-0.0211(22)	-0.0215(25)	0
	z	0.1815(3)	0.1814(3)	0.1819(3)
	B	0.3(2)	0.2(3)	0.7(1)
	Occupancy	0.94(2)	0.94(2)	0.91(2)
O(2)	z	0.0043(8)	0.0047(9)	0
	B	0.3(2)	0.03(2)	0.3(1)
	Occupancy	0.96(2)	0.92(2)	0.89(2)
χ^2		0.991	0.975	1.236

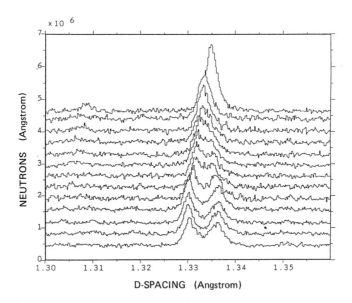

Fig. 5 The (040, 400) peaks in the powder neutron diffraction of $La_{1.85}Sr_{0.15}CuO_4$ from (top) 300K to (bottom) 22K.

are available for enough temperatures to provide a rough estimate of the critical exponent β and the tetragonal–orthorhombic transition temperature T_d according to the equation

$$(a-b)/(a+b) \propto \left[(T_d-T)/T_d\right]^{-\beta} \qquad (1)$$

We find β=0.5(2) and T_d=180(20)K; the value of β is close to that predicted[11] by mean field theory, no doubt because the data do not extend below 0.05 in reduced temperature. A comparable value has been found for the I4/mmm → Abcm transition in $La_{1.8}Ba_{0.2}CuO_4$. (12)

   The maximum distortion, measured by the dimensionless units $(a-b)/(a+b)$, is extremely small (2.3×10^{-3}). It occurs between 30 and 40K and then appears to diminish again by an amount that exceeds the error bar of the measurement (Fig. 4(a)), as does the c-axis parameter (Fig. 4(b)). This unusual behaviour should be compared with that of the corresponding La,Ba superconductor, (Paul et al. 1987) which is also tetragonal at room temperature. An orthorhombic distortion also sets in at 177(2)K, again increasing smoothly with decreasing temperature with β=0.56(3), though to a smaller maximum value of $(a-b)/(a+b) = 1.4 \times 10^{-3}$. However, there is a crucial difference between the two compounds because in the Ba one the orthorhombic distortion starts to decrease again at about 80K, reaching a smaller constant, but non-zero, value at about 50K. Both compounds have superconducting T_c's between 35 and 40K, but in the Sr compound, the maximum orthorhombic distortion is larger, and it begins to decrease at a much lower temperature. Most importantly, there is no observable discontinuity in the temperature variation of the crystal structure of $La_{1.85}Sr_{0.15}CuO_4$ near the superconducting transition. For example, within the limitation of the error bars, the variation of the Cu-O(2)-Cu angle with temperature is smooth (Fig. 6).

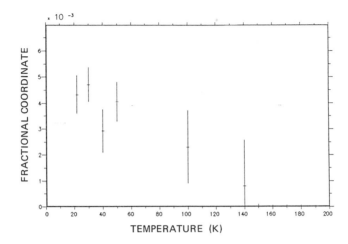

Fig. 6   Fractional z coordinate of O(2) as a function of
         temperature.

The most important conclusion of all from the neutron
diffraction experiments on both the La,Sr and La,Ba compounds is
the absence of any direct correlation between the superconducting
transition temperature and that of the tetragonal-orthorhombic
structural transition.   It has been proposed that doping La_2CuO_4
with the Group 2 cations Sr or Ba suppresses the $2k_F$ instability
which would open a gap at the Fermi surface if the $\sigma^*(x^2-y^2)$ band
was exactly half-filled.   The Fermi energy decreases as the
average formal Cu oxidation state rises, and an early suggestion
was that the electron-phonon coupling arises from low frequency
librational modes associated with the tilting of the $CuO_6(13)$.
However,   the   experiments   described   above   show   that
superconductivity   is   only   found   in $La_{1.85}Sr_{0.15}CuO_4$ at   a
temperature well below that at which the orthorhombic distortion
sets in.   In   other   words,   superconducting $La_{1.85}Sr_{0.15}CuO_4$
has the same orthorhombic symmetry as semiconducting La_2CuO_4.

4.   VIBRATIONAL SPECTRUM

Although there is no detectable oxygen isotope effect on T_c of the 90K superconductor $YBa_2Cu_3O_/$ (14), it is reported (15) that such an effect does exist in the 40K materials, albeit smaller than found in prototypic superconductors like Hg and Pb which obey BCS theory. The presence of even a small isotope effect requires at least some participation by lattice vibrations in the formation of the current-carrying Cooper pairs, and, if one believes that the mechanisms in the 40K and 90K series are similar, implies that phonons cannot be neglected in the latter. Thus the lattice vibrational spectrum assumes a great importance.

There are two reasons for turning to inelastic neutron scattering to measure the vibrational spectrum: first, the compounds concerned are metallic, so direct measurement of infrared or Raman spectra is rendered difficult by their opacity; second, the various kinds of photon spectroscopy only provide information about the energies of modes close to the Brillouin zone centre. The lower energy modes of continuous lattice compounds such as these, especially the acoustic modes of greatest importance in the BCS theory, show a strong energy dispersion, so the zone-centre frequency alone is not informative enough. Ideally, phonon dispersion curves are measured by coherent inelastic neutron scattering using a triple-axis spectrometer, but unfortunately such measurements require large ($1cm^3$) single crystals. In cases when only powders are available, directional information is lost, though the integrated phonon density-of-states (DOS) can be obtained from incoherent inelastic neutron scattering (IINS).

In view of the suggestion (13) that vibrational modes associated with the tilting of the CuO_6 octahedra might be the source of the strong electron-phonon interaction in $La_{1.85}Sr_{0.15}CuO_4$, we chose to concentrate on the low frequency part of the vibrational spectrum. (Alternative suggestions (16)

have pointed to the in-plane Cu-O(2) modes, discussed in Section
5). The experiments described here (17) were performed on the IN6
time-of-flight spectrometer at the Institut Laue-Langevin,
Grenoble, a high flux research reactor operated by France, Germany
and the U.K. IN6 is on a cold neutron guide, providing 5.159Å
neutrons from graphite monochromators. After being collimated and
chopped, the beam is scattered from the sample, which is in the
form of a flat disc about 5cm in diameter and a few mm thick,
sandwiched between aluminium foil. The scattered neutrons
traverse a helium filled flight path of 3m and impinge on 337 high
pressure ^{3}He detectors. IN6 works in neutron energy gain, i.e. by
transferring energy from a thermally populated vibrational mode to
the outgoing neutron. It is therefore at its best at high
temperatures and low energy transfers. Absolute scattering
cross-sections are derived from the raw data by comparison with a
vanadium standard to give the $\underline{Q}$-dependent frequency distribution
P(A,B) which is then extrapolated to zero momentum transfer to
yield the amplitude-weighted DOS, P(ω).

Figure 7 shows P(ω) for $La_{1.85}Sr_{0.15}CuO_4$ and La_2CuO_4
measured near room temperature from 0-100 meV (1 meV $\sim$ $8cm^{-1}$).
The worsening statistics with increasing energy transfer due to
the fall off in thermal phonon population are clearly apparent,
but features in the profile are clearly discernible up to 80meV.
Both DOS plots are highly structured, and are most notable for
their striking similarity. Indeed, there are no peaks in P(ω) of
the superconductor which are not present in La_2CuO_4, and at very
similar energies. Peaks are found at the following energies:
La_2CuO_4 : 12   18.5   (24) 29.5 39.8 57.2   80   meV
$La_{1.85}Sr_{0.15}CuO_4$:   11.5 18.5   23.8 29.5 38   55.2   78
The lowest energy peak, assigned to the zone-edge acoustic mode,
has been observed with much poorer resolution and statistics in an
experiment on a triple-axis spectrometer at Brookhaven using 60g
of powdered $La_{1.85}Sr_{0.15}CuO_4$[17], and the 20meV peak,
again with very poor signal-to-noise, on the High Energy Transfer

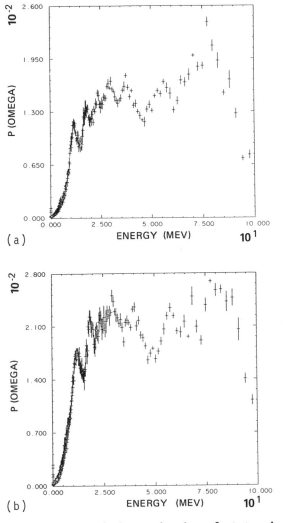

Fig. 7   Amplitude weighted phonon density of states in (a) La$_{1.85}$Sr$_{0.15}$CuO$_4$ at 250K, (b) La$_2$CuO$_4$ at 300K.

Spectrometer at the ISIS pulsed neutron source (18). It has been suggested (17) that the 12meV peak is the cause of an anomaly observed in the heat capacity at low temperature.

The close similarity between Fig. 7(a) and 7(b) is also important in connection with the structural distortion and possible presence of soft modes because at the temperatures at which the $P(\omega)$ were measured, the superconducting phase is above the tetragonal-to-orthorhombic transition while La_2CuO_4 is well below it. Lowering the symmetry of the La compound has not introduced any notable new features into its DOS, compared with the tetragonal La,Sr compound. The same is also true of the latter, when we compare the low energy part of the DOS at temperatures above and below T_d. In Fig. 8(a) and (b) we show $P(A,B)$ curves for $La_{1.85}Sr_{0.15}CuO_4$ up to 20meV at 200 and 100K. They have been normalized to take account of the difference in thermal populations between the two temperatures. Straightaway one observes that they are essentially identical. There is apparently no softening of any low energy mode up to 20meV $(160cm^{-1})$ on passing through T_d. Also worth noticing is the classical ω^2 dependence of the DOS below 8meV, corresponding to a single dominating Debye frequency.

An enlargement of the same energy region of the $P(A,B)$ of La_2CuO_4 at 100K (Fig. 9) contains broadly comparable features but, by comparison with $La_{1.85}Sr_{0.15}CuO_4$, increased scattering below about 8meV, apparent as a broadening of the 12meV peak. Because of this, the ω^2 law is only obeyed up to about 6meV. According to susceptibility measurements[4], at 100K La_2CuO_4 is well below the temperature (250K) at which a transition occurs to a spin density wave or antiferromagnetic state. Hence it is possible that the increased low energy neutron scattering intensity in this compound may arise from magnetic excitations (magnons). By examining the intensity from different banks of detectors, we also have indications that this scattering is $\underline{Q}$ dependent, as expected for magnetic scattering because of the $\underline{Q}$ dependence of the electronic form factor. Confirmation of this hypothesis must await inelastic neutron scattering measurements with polarisation analysis, which will distinguish phonon (non spin flip) from magnon (spin flip) scattering.

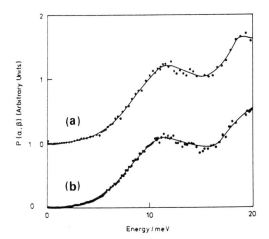

Fig. 8   Phonon density of states in La$_{1.85}$Sr$_{0.15}$CuO$_4$ up to
20 meV at (a) 200 and (b) 100K, corrected for change of
phonon population.

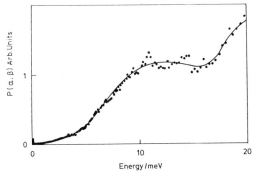

Fig. 9   Phonon density of states in La$_2$CuO$_4$ at 100K.

5.   MECHANISMS

What do the experiments described in Sections 2 and 3 tell us
about the mechanism of superconductivity in the remarkable Cu
oxide ceramics?   Before answering this question, it is worth
summarizing some of the other facts about them.

Quantized flux expulsion experiments (19) prove that the
current carriers are electron pairs (Cooper pairs), as in conventional superconductors.   They are type II superconductors, with

extremely high critical fields B_{c2} (20), from whose temperature dependence in the neighbourhood of T_c, estimates of the coherence lengths ξ of about 20Å have been obtained (21). Such values are much smaller than in simple metals like $Al(10^4 Å)$ or A15 compounds(50Å), and of the order of magnitude of the unit cell parameters. Apart from the higher T_c, comparable values of B_{c2} and ξ are found for $M^{III}Ba_2Cu_3O_{7-\delta}$, whose structures likewise contain layers of corner-sharing square CuO_4 units. Indeed, the latter feature is the only point of structural similarity between the 40K and 90K series: in the 40K compounds the CuO_4 form part of the strongly elongated octahedral CuO_6 and in the 90K compounds part of the square pyramidal CuO_5.(5) In addition, chains of corner-sharing CuO_4, with no additional axially coordinated O, lie parallel to the b-axis in $YBa_2Cu_3O_{7-\delta}$. No comparable structural motif exists in $La_{2-x}M_x^{II}CuO_{4-y}$, so it seems reasonable to conclude that it is within the layers of corner-sharing CuO_4 that we should seek the origin of the superconductivity, while recognizing that the CuO_4 chains may serve to enhance T_c.

A second common feature between the two groups of compounds is the presence of a small descent in symmetry from tetragonal to orthorhombic. At first it was thought that, while La_2CuO_4 had a small orthorhombic distortion from the tetragonal K_2NiF_4 structure it was rendered tetragonal on doping with Sr or Ba.(13). However, we now know that both $La_{1.85}M_{0.15}CuO_4$ (M=Sr,Ba) undergo second-order transitions from tetragonal to orthorhombic near 170K (12,22), well above the temperatures (respectively about 40 and 35K) at which they become superconducting. Similarly, $MBa_2Cu_3O_{7-y}$, first thought to be tetragonal, with disordered anion vacancies (23), is now known to be orthorhombic (Pmmm) with ordered vacancies (5).

The orthorhombic distortion of CuO_2 layers in La,Sr(Ba) results from slight puckering, so that along the a-axis the oxide ions are alternately above and below the planes of the Cu. Since the orthorhombic-to-tetragonal transition is second order (12,22) the Cu-O-Cu angle varies with temperature (Fig. 6), but in La,Sr it reaches a limiting value of 174° at 30K and in La,Ba 176° at about 80K. In La$_{2-x}$Sr$_x$CuO$_4$ the maximum orthorhombic distortion and T$_d$ decrease with increasing x till at x=0.20 the tetragonal phase is stable down to T=0 (24). This is just the composition beyond which superconducivity is no longer found.

In YBa$_2$Cu$_3$O$_{7-\delta}$ the origin of the orthorhombic distortion is the ribbons of corner sharing square planar CuO$_4$ groups parallel to the b-axis. The unshared corners of the latter CuO$_4$ form the apices of the square pyramidal CuO$_5$ that constitute the two-dimensional CuO$_2$ array in this lattice type. In this compound, too, there is a high temperature tetragonal phase, formed by disordering the anion vacancies, and as in the case of the La$_{2-x}$M$_x^{II}$CuO$_4$ compounds, it is not superconducting.

If an orthorhombically distorted CuO$_2$ layer, not differing much in energy from a parent tetragonal layer, is a common feature of the two sets of high T$_c$ superconductors, it cannot be because of a high density of low energy phonon modes associated with torsion of the CuO$_4$ units, since our IINS measurements show no trace of such modes. The low energy part of the phonon spectrum of La$_{1.85}$Sr$_{0.15}$CuO$_4$ is quite unaffected by the tetragonal-orthorhombic transition at 170K. Are a different set of modes instrumental in forming the Cooper pairs?

Given the short correlation length, the electron pair Bosons are close to being formed in real, as opposed to momentum space. Superlattices of spatially localized electron pairs (called by physicists 'charge density waves' CDW) are a common feature of mixed valency compounds of the elements in Groups 10-15 of the Periodic Table (25). Classical examples are Wolfram's Red Salt $[Pt^{II}(C_2H_5NH_2)_4][Pt^{IV}(C_2H_5NH_2)_4Cl_2]Cl_4$, Wells' Salt $Cs_2[Au^ICl_2][Au^{III}Cl_4]$, $Cs_4[Sb^{III}Cl_6][Sb^VCl_6]$ and $Ba_2Bi^{III}Bi^VO_6$. The last of these is the parent compound of the relatively high temperature(T$_c$=14K) superconductors BaBi$_{1-x}$Pb$_x$O$_3$, while Wells' Salt has particular significance in the context of the Cu superconductors. Not only does the metal ion originate from the same Periodic Group, with a similar mean electron configuration d^9, but its crystal structure perpendicular to its tetragonal c-axis consists of layers of corner-sharing $AuCl_4$ square units, precisely as in the Cu superconductors. There is a crucial, and illuminating, difference, however, because the CuO_4 in the basal plane are all the same while in Wells' Salt half the $AuCl_4$ have much shorter Au-Cl distances than the other half (26). In other words the average d^9 configuration has 'disproportionated' to d^8 and d^{10}.

To pair electrons in a single orbital costs energy, so the existence of a ground state $e^2e^0e^2$ .....rather than $e^1e^1e^1$ implies that the electrons are "attracted" together. The distortion of the lattice produces this apparent attraction, overcoming the actual repulsion, just as in superconductors. If the electron-phonon interaction is strong, or the single site electron repulsion weak, we find a static CDW superlattice of electron pairs as in Wells' Salt. The vibronic theory for discrete dimeric metal complexes of this type has recently been given[27], and the single site electron repulsion estimated for a Sb(III,V) system (28), which is a close CDW analogue of the $BaBi_{1-x}Pb_xO_3$ superconductors. The 5d orbitals in Wells' Salt are larger than the 3d ones in the Cu oxide superconductors, and so electrons within them pair more easily. Also the electron band width is larger in the Cu oxides. Possibly the pairs carrying the current in the latter compounds are the fluctuating analogues (bipolarons (29)) of the static pairs in Wells' Salt. Furthermore, it has been shown recently that for a large mean phonon energy, relative to the electron-phonon coupling energy, the superconducting energy gap for a two-dimensional N-site N-electron array of the kind found in the Cu oxide superconductors is independent of changes in the phonon energy (30), thus explaining the weak isotope effect. No anomalous or temperature dependent phonon DOS would need to be invoked or looked for, and the function of the orthorhombic distortion would be to ensure that the Fermi surface did not fall at a singularity in the electronic DOS. The preparation of other compounds based on these hypotheses would serve to confirm them.

ACKNOWLEDGMENTS

The experiments described in this chapter have been done with enthusiastic help from the following: Matthew Rosseinsky and Mohamed Kurmoo (Inorganic Chemistry Laboratory, Oxford), Harry Jones and John Singleton (Clarendon Laboratory, Oxford), Kosmas Prassides (formerly ICL, Oxford now University of Crete), Bill David, Oscar Moze and Alan Soper (Rutherford Appleton Laboratory) and A.J. Dianoux (ILL Grenoble).   Their names appear in the references.   We acknowledge support of SERC, and the Directors of RAL and ILL for access to neutron beams.

REFERENCES

1. D.W. Capone, D.G. Hinks, J.D. Jorgensen and K. Zhang, Appl. Phys. Lett. 50, 543 (1987); R.J. Cava, R.B. Van Dover, B. Batlogg and E.A. Rietmann, Phys. Rev. Lett. 58, 408 (1987).

2. R.J. Cava, B. Batlogg, R.B. Van Dover, D.W. Murphy, S. Sunshine, T. Siegrist, J.P. Remeika, E.A. Rietman, S. Zaharak and G.P. Espinosa, Phys. Rev. Lett. 58, 1676 (1987).

3. A.J. Panson, A.I. Braginski, J.R. Gavaler, J.K. Hulme, M.A. Janocko, H.C. Pohl, A.M. Stewart, J. Talvacchio and G.R. Wagner, Phys. Rev. B35, 8774 (1987); J.D. Jorgensen et al., Phys. Rev. B, submitted; A. Santoro et al., Mat. Res. Bull., submitted.

4. R.L. Greene, H. Maletta, T.S. Plaskett, J.G. Bednorz and K.A. Muller, Sol. St. Commun. in press.

5.   W.I.F. David, W.T.A. Harrison, J.M.F. Gunn, O. Moze, A.
     Soper, P. Day, J.D. Jorgensen, M.A. Beno, D.W. Capone, D.C.
     Hinks, I.K. Schuttler, L. Soderholm, C.V. Segre, K. Zhang and
     J.D. Grace, Nature 327, 310 (1987); J.J. Capponi, C.
     Chaillot, A.W. Hewat, P. Lejay, M. Marezio, N. Nguyen, B.
     Raveau, J.L. Soubeyroux, J.L. Tholence and R. Tournier,
     Europhys. Lett. 3, 301 (1987).

6.   J.G. Bednorz and K.A. Muller, Z. Phys. B64, 189 (1986).

7.   H. Takagi, S. Uchida, K. Kitazawa and S. Tanaka, Jap. J.
     Appl. Phys. Lett. 26, 41 (1987).

8a. H.H. Wang et al., Inorg. Chem. 26, 1474 (1987); (b) C.
     Marailly, P. Courty and B. Delmon, J. Am. Ceram. Soc. 53, 56
     (1970).

9a. R.J.D. Tilley and A.H. Davies, Nature 326, 859 (1987); (b)
     J.M. Longo and P.M. Raccah, J. Sol. St. Chem. 6, 526 (1973);
     V.B. Grande, H. Muller-Buschbaum and M. Schweizer, Z. Anorg.
     Allg. Chem. 428, 120 (1977).

10. H.M. Rietveldt, J. Appl. Cryst. 20, 508 (1969).

11. Kittel, Introduction to Solid State Physics, New York, John
     Wiley & Son.

12. D. McK. Paul, G. Balakrishnan, N. Bernhoeft, W.I.F. David and
     W.T.A. Harrison, Phys. Rev. Lett. 58, 1976 (1987).

13. J.D. Jorgensen, J.B. Schuttler, D.G. Hinks, D.W. Capone, K.
     Zhang, M..B. Brodsky and D.J. Scalopino, Phys. Rev. Lett. 58,
     1024 (1987).

14. B. Batlogg, R.J. Cava, A. Jayuraman, R.B. van Dover, G.A.
     Kourouklis, S. Sunshine, D.W. Murphy, L.W. Rupp, H.S. Chen,
     Whatt, K.T. Short, A.M. Mujsee and E.A. Rietman, Phys. Rev.
     Lett., in press.

15. Reported by Berkeley and Bell groups at Adriatica Conference,
     Trieste, July 1987.

16. W. Weber, Phys. Rev. Lett. 58, 1371 (1987).

17. M. Rosseinsky, K. Prassides, P. Day and A.J. Dianoux, Phys.
     Rev. B, submitted

18. A.P. Ramirez, B. Batlogg, G. Aeppli, R.J. Cava, E. Reitman, A. Goldman and G. Shirane, Phys. Rev. B35, 8833 (1987).

19. D. McK. Paul, G. Balakrishnan, N.R. Bernhoeft and A.D. Taylor, Nature 327, 45 (1987).

20. C. Gough et al, Nature, 326, 855 (1987).

21. S. Vehida, H. Tagaki, S. Tanaka, K. Nakao, N. Mirra, K. Kishio, K. Kitazaka and K. Fueki, Jap. J. Appl. Phys. Lett., preprint; ref (2).

22. T.K. Worthington et al., Proc. Workshop on Novel Mechanisms in Superconductivity, Berkeley, June 1987, in press.

23. P. Day, M. Rosseinsky, K. Prassides, W.I.F. David, O. Moze and A. Soper, J. Phys. C.: Sol St. Phys. 20, L429 (1987).

24. R.M. Hazen, L.W. Finger, R.J. Augel, C.T. Prewitt, N.L. Ross, H.K. Mao, C.G. Hadidiacos, P.H. Hor, R.L. Meng and C.W. Chu, Phys. Rev. B35 7328 (1987); H. Steinfink, J.S. Swinnen, Z.T. Sui, H.M. Hsu and J.B. Goodenough, J. Amer. Chem. Soc. 109, 3348 (1987).

25. R. Monet, J.P. Pouget and G. Collin, Europhys. Lett., 1 August 1987.

26. M.B. Robin and P. Day, Adv. Inorg. Chem. & Radiochem. 10, 247 (1967); P. Day, Int. Rev. Phys. Chem., 1, 149 (1981).

27. N. Elliott and L. Pauling, J. Amer. Chem. Soc. 60, 1846 (1938).

28. K. Prassides, P.N. Schatz, K.Y. Wong and P. Day, Phys. Chem. 90, 5588 (1986).

29. K. Prassides and P. Day, Inorg. Chem. 24, 1109 (1985).

30. A. Alexandrov, J. Ranninger and S. Robuszkiewicz, Phys. Rev. B33, 4526 (1986).

31. K. Nasu, Phys. Rev. B35 1748 (1987).

15

Observation of the AC Josephson Effect Inside Copper Oxide-Based Superconductors

M. H. DEVORET, D. ESTEVE, J. MARTINIS, C. URBINA
Service de Physique du Solide et de Résonance Magnétique,
CEN-Saclay 91191 GIF SUR YVETTE CEDEX (France)

G. COLLIN, P. MONOD and M. RIBAULT Laboratoire de Physique
des Solides, Université Paris Sud 91406 ORSAY (France)

A. REVCOLEVSCHI, Laboratoire de Chimie des Solides
Université Paris Sud, 91406 ORSAY (France)

INTRODUCTION

In past research on superconductivity, tunneling experiments proved to be useful tools in the investigation of its microscopic origin (1). They led to the experimental demonstration of the Josephson effect (2) which is a direct manifestation of the quantum phase of the superconductive order parameter and of the coupling of this phase to the electromagnetic field. In particular, the ac Josephson effect links the time evolution of the phase difference between two superconductors to the voltage between them. When this voltage is kept fixed externally, the phase difference increases linearly with time, giving rise to an oscillating supercurrent between the superconductors if they are coupled by a weak link such as a tunnel barrier. This effect depends only on the bose-like character of the particles carrying the supercurrent and it has an equivalent for the superfluids ^{4}He and ^{3}He which has recently been observed (3). The relation between voltage V and frequency f is simply given by $V = (h/q)f$ where h is Planck's constant and q the charge of the bose-like supercurrent carrier. In usual superconductors, the observation of the ac Josephson effect yields $q = 2e$ (4). Historically, this provided a further confirmation (after the discovery of flux quantization (5)) of the existence of Cooper pairs, but other values of q are theoretically possible (6).

It was thus important to test if the Josephson effect could be observed with copper oxide based superconductors (7). This was achieved by our experiment performed at Saclay (8) and by other experiments conducted independently (9). Since the details of our experiment have been reported elsewhere (8), we will focus in this paper on our latest results and their discussion.

EXPERIMENTAL PROCEDURES AND RESULTS

Our samples of $La_{1.85}Sr_{0.15}CuO_4$ and $Y_1Ba_2Cu_3O_{7-x}$ were mounted in a point contact (10) current-voltage (I-V) probe which consists of a copper sample holder and a conducting tip whose pressure onto the sample could be remotely ajusted by means of a screw. Different metals were used for the tip : Nb, Cu, Al, Fe, Au, brass and copper-berryllium. The I-V characteristic of the tip + sample + holder system was measured using a three wire arrangement with filters to prevent ambiant noise from reaching the sample. The series resistance of the tip and holder always played a negligible role in the measurements. The temperature of the probe could be

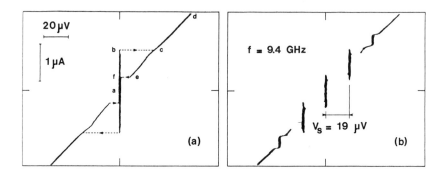

FIGURE 1 a) Oscilloscope trace of a current-voltage characteristic obtained at 4.2 K with an aluminium tip on a $La_{1.85}Sr_{0.15}CuO_4$ sample. Letters a through f indicate sense of trace. Dashed lines indicate the switching between the branches. b) Steps induced by microwave irradiation.

varied continuously from room temperature down to 4.2 K and was
monitored by a simple nickel/manganin thin film resistor.
Microwaves of frequency varying between 1 and 18 GHz were coupled
capacitively to the junction.

At liquid helium temperature we found that, irrespective of
the metal used for the tip, we could always obtain hysteretic I-V
characteristics of the type shown on Fig. 1a obtained with a
$La_{1.85}Sr_{0.15}CuO_4$ sample. These characteristics consist of two
branches. Increasing the current from zero (point a), one follows
first a "zero-voltage" branch up to a point (b) where one switches
to a "resistive" branch (c ↦ d). Upon decreasing the current one
follows the resistive branch down to a second switching point (e)
where one returns to the zero voltage branch (f). This cycle
repeats itself symmetrically for negative currents. We found it
necessary to move the tip away from the sample and to re-ajust the
pressure several times before finding a good zero-voltage branch -
i.e. with an apparent resistance markedly lower than that of the
resistive branch. Presumably, this motion allows the selection of a
favorable location on the sample with minimal series resistance,
since the exact point probed by the tip varied randomly as the tip
was moved away from and back onto the sample. Fig. 2a shows an I-V

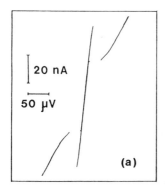
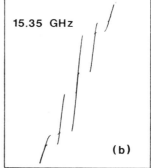

FIGURE 2 a) Current voltage characteristic of a $Y_1Ba_2Cu_3O_{7-x}$ sample
at 4.2 K with an aluminium tip. b) Steps induced by microwave
irradiation.

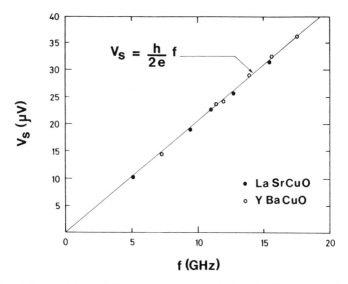

FIGURE 3 Separation between microwave induced steps as a function
of frequency. Solid line is the prediction for a superconductor
with flux quantum h/2e.

characteristic of a $Y_1Ba_2Cu_3O_{7-x}$ sample. The series resistance for
this material was systematically larger than for the
$La_{1.85}Sr_{0.15}CuO_4$ samples. We define the critical current I_c as the
current for which the switching b ↦ c occurs. Both the critical
current and the amount of hysteresis are pressure and location
dependent quantities.

The shape of the I-V characteristic does not by itself
demonstrates its Josephson junction origin. A convincing
confirmation of this origin, however, is provided by the result of
microwave irradiation. Well defined current steps equidistant in
voltage are then induced on the I-V characteristic (see Figs. 1b
and 2b). These Shapiro steps (4) result from the beating of the
oscillating Josephson supercurrent with applied microwave
radiation. Steps of rank up to 5 were observed. In Fig. 3 we plot
the separation V_s between two consecutive steps as a function of
the microwave frequency f. We see that within the experimental

accuracy, the data obey the ac Josephson relation $V_s = (h/2e)f$. The same value of the flux quantum $h/2e$ is found in other ac Josephson effect experiments (9) and in flux quantization measurements (11).

A further confirmation of the attribution of the steps of Figs. 1b and 2b to the ac Josephson effect is obtained when one varies the microwave amplitude at a fixed frequency. We find that the current amplitude I_n of the step of rank n is modulated in a Bessel function-like pattern. Bessel functions are expected when two superconductors are coupled by a weak link providing a sinusoidal current-phase relation (dc Josephson relation) (10).

DISCUSSION

It is important to note that the characteristics of Figs. 1 and 2 were observed with a non-superconducting - and in one instance magnetic - metal tip. Furthermore, in several instances, the characteristics showed two or three junctions in series. We thus conclude that the characteristics originates from a junction inside the sample rather than between tip and sample as in usual point contacts. The junctions, which we think are located near the tip where the current density is the highest, may result from cracks, grain boundaries or other defects of the samples. The presence in similar materials, even in single crystal form, of junctions which would form a random network is now well established and is thought to be due to the very short coherence length of these new superconductors (12).

In order to measure the dimension that would characterize such a network, a magnetic field was applied parallel to the tip by means of a superconducting coil wound on the copper shield surrounding the probe. The critical current of the junctions always showed a marked dependence on magnetic field and in several cases we observed an oscillatory pattern reminiscent of the diffraction pattern of a dc SQUID (10). The first minimum of I_c occured typically at B = 100 Oe. If we assume that in this case the I-V characteristic results from two junctions in a superconducting loop, we find an area of order 0.1 μm^2 for the projection of the loop perpendicular to the magnetic field. Taking this loop as representative of loops in a random network of junctions, this order of magnitude is consistent with the interpretation of

Deutscher and Müller (12).

It is worth noting that all the I-V characteristics we observed became increasingly rounded as the temperature was raised. For a given sample, characteristics of junctions with higher critical current tend to survive at higher temperatures than those with lower critical currents. Typically, in the Sr doped La_2CuO_4 samples, characteristics with I_c = 1 µA become completely rounded at T = 6 K - i.e. well before reaching the superconductive temperature. However, in one $Y_1Ba_2CuO_{7-x}$ sample, we could follow a junction with I_c = 50 nA up to 50 K. We have no quantitative explanation for the evolution of the I-V characteristics with temperature but similar rounding is observed in standard Josephson junctions where it is due to a combination of dissipation and thermal fluctuations of the phase difference of the junction (10). This rounding seems to depend on the way the junctions are prepared since Tsai et al (13) and Moreland et al (14) have been able to follow Shapiro steps up to 85 K using the break junction technique.

In conclusion, we have observed I-V characteristics of Josephson junctions inside copper oxide based superconductors. Even though the exact location in the samples and the origin of the junctions remained unknown, we could use them to demonstrate that the phase of the superconductive order parameter is coupled to the electromagnetic field through the fundamental constant h/2e. Our results directly imply that in these newly discovered materials, like in conventional superconductors, electron pairing is responsible for the superconductivity.

ACKNOWLEDGMENTS

We have benefited from interesting and stimulating discussions with K. A. Müller and Prof. R. A. Buhrman. Help in sample preparation from J. Jegoudez and technical assistance by C. Noël are gratefully acknowledged. We are also indebted to P. Tremblay for the analysis of samples.

REFERENCES

1. I. Giaver, Phys. Rev. Lett., 5, 147, 464 (1960).
2. B. D. Josephson in Superconductivity (R. Parks ed.), Marcel Dekker, New York , 1969, pp. 423–448.
3. O. Avenel and E. Varoquaux, Proceedings of LT18 (Kyoto) (1987).
4. S. Shapiro, Phys. Rev. Lett., 11, 80 (1963).
5. B. S. Deaver and W. M. Fairbank, Phys. Rev. Lett., 7, 43 (1961); R. Doll and M. Näbauer, Phys. Rev. Lett., 7, 51 (1961)
6. M. H. Cohen and D. H. Douglass, Phys. Rev. B35, 8720 (1987); S. A. Kivelson, D. S. Rokhsar, and J. P. Sethna, Phys. Rev. B35, 8865 (1987).
7. J. G. Bednorz and K. A. Müller, Z. Phys. B, 64, 189 (1986); M. K. Wu et al, Phys. Rev. Lett., 58, 908 (1987).
8. D. Esteve et al, Europhys. Lett., 3, 1237 (1987).
9. J. S. Tsai, Y. Kubo and J. Tabuchi, Phys. Rev. Lett., 58, 1979 (1987); W. R. McGrath et al, Europhys. Lett., 4, 357 (1987); J. Kuznik et al, preprint submitted to J. Low Temp. Phys.
10. A. Barone and G. Paterno, Physics and Applications of the Josephson Effect, John Wiley, New York, 1982.
11. C. E. Gough et al, Nature, 326, 855 (1987).
12. G. Deutscher and K. A. Müller, preprint submitted to Phys. Rev. Lett..
13. J. S. Tsai, Y. Kubo and J. Tabuchi, Jpn. J. Appl. Phys., 26, L701 (1987).
14. J. Moreland et al, preprint submitted to Appl. Phys. Lett..

16

Temperature-Dependent Conductivity of Oxygen-Depleted YBCO Ceramics

J. H. MILLER, JR.,   B. LIU,   W. J. RILEY,   A. N. DIBIANCA,   S. L.
HOLDER,   and J. D. DUNN   Department of Physics and Astronomy,
University of North Carolina, Chapel Hill, North Carolina 27599-3255

B. R. ROHRS   and W. E. HATFIELD   Department of Chemistry, University
of North Carolina, Chapel Hill, North Carolina 27599-3290

Considerable effort has been devoted to characterizing the properties
of the recently discovered copper-oxide superconductors.[1-7] However,
there have been relatively few systematic studies[8-10] reporting on
the electrical transport properties of nonsuperconducting copper-
oxide perovskites, despite the fact that nonsuperconducting materials
comprise a much larger phase space than the superconducting
specimens. Relatively minor changes in oxygen content are sufficient
to convert a superconducting "123" compound into a nonsuper-
conductor. For example, $YBa_2Cu_3O_{6.9}$ is a superconductor, whereas
$YBa_2Cu_3O_{6.4}$ is not. We suggest that the term "incipient
superconductor" be applied to compounds such as $YBa_2Cu_3O_{6.4}$ and
La_2CuO_4, since simply annealing the former in oxygen or doping the
latter with an alkaline earth is sufficient to convert the respective
compound into a high T_c superconductor.

We believe that it is important to systematically characterize
the behavior of nonsuperconducting copper-oxide perovskites for

several reasons: 1) it has been demonstrated[11] that damage by ion implantation converts perovskite thin films into semiconductors, suggesting the possibility of developing superconductor/ semiconductor hybrid circuits for microelectronic applications, 2) a detailed understanding of the known incipient superconductors may shed light on the pairing mechanism and the conditions necessary for high T_c superconductivity, and 3) the search for new classes of high temperature superconducting materials will be aided considerably if signatures of incipient superconductivity are identified.

We have therefore initiated a study of nonsuperconducting copper-oxide perovskites, beginning with oxygen-depleted "123" ceramics. Figures 1 and 2 show plots of the temperature-dependent resistances of ceramics of $YBa_2Cu_3O_{6.9}$ (orthorhombic phase) and $YBa_2Cu_3O_{6.4}$ (tetragonal phase), respectively. Note that the

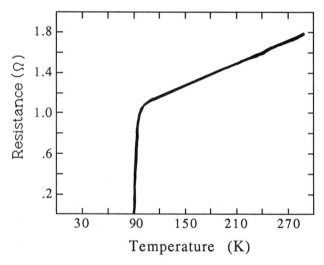

Figure 1. Temperature-dependent resistance of $YBa_2Cu_3O_{6.9}$.

resistance of $YBa_2Cu_3O_{6.4}$ changes little from room temperature down to around 200 K, below which the resistance increases rapidly with decreasing temperature. This suggests that a metal-insulator transition may be occuring near 200 K. Figure 3 shows an Arrhenius plot of the temperature-dependent conductivity of the oxygen-depleted

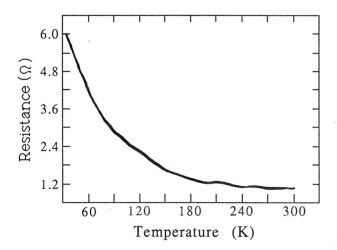

Figure 2. Temperature-dependent resistance of $YBa_2Cu_3O_{6.4}$.

sample, and indicates that activated behavior,

$$\sigma \simeq \sigma_0 \exp[-\Delta/k_B T], \qquad (1)$$

is only approximated in the temperature range 120 K < T < 200 K, if at all. In this region the activation energy is estimated to be Δ/k_B = 165 K, which is equal to the BCS gap parameter Δ/k_B = 1.76 T_c that one would obtain for a superconductor with a transition temperature of T_c = 93 K in the weak coupling limit.

The effects of disorder are likely to play an important role in the electrical transport properties of oxygen-depleted samples, because of the random distribution of oxygen vacancies. Sufficient disorder often leads to a high density of localized states, so that the conductivity may become dominated by hopping processes. For the

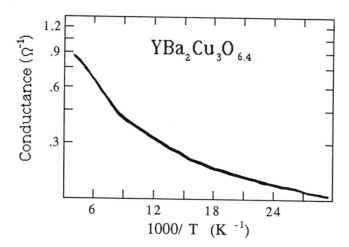

Figure 3. Arrhenius plot of the temperature-dependent conductance of $YBa_2Cu_3O_{6.4}$.

case of variable range hopping, the conductivity is predicted to have the form[12]

$$\sigma = \sigma_0 \exp\left(-A/T^{\frac{1}{4}}\right) \tag{2}$$

where

$$A = 2 \left(\frac{3}{2\pi}\right)^{\frac{1}{4}} \left\{\frac{\alpha^3}{k_B N(E_F)}\right\}^{\frac{1}{4}} \tag{3}$$

and $1/\alpha$ is the average decay length of the localized wave functions. Figure 4 shows a plot of conductivity vs. $T^{-\frac{1}{4}}$ on a semilog scale, and indicates that Eqn. (2) describes the conductivity in the temperature range 70 K < T < 190 K significantly better than Eqn. (1). In this temperature range, the constant A is inferred to be A ~ 13.7 $K^{\frac{1}{4}}$, so that the characteristic "hopping energy" is given by $\alpha^3/N(E_F)$ ~ 0.4 eV. Similar behavior has been observed in single crystal $La_2Cu_3O_4$,[8] which is believed to undergo an antiferro-magnetic phase transition.

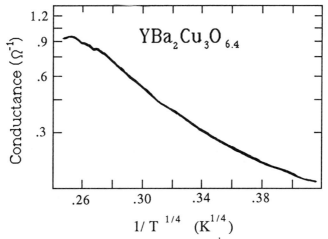

Figure 4. Plot of conductance *versus* $1/T^{\frac{1}{4}}$ for $YBa_2Cu_3O_{6.4}$, showing evidence for variable-range hopping between localized states.

To summarize, we have reported on measurements of the temperature-dependent conductance of oxygen-depleted YBCO ceramics, and find that the conductance seems to be more accurately described as variable range hopping between localized states than as activated conductivity characteristic of a semiconductor. A detailed, systematic study of a wide range of nonsuperconducting copper-oxide perovskites is likely to yield important information and is currently in progress.

ACKNOWLEDGMENTS

We wish to thank A. Kingon and D. Haase for useful, as well as stimulating, conversations. This research was supported in part by the Office of Naval Research. Equipment support was provided by the

Department of Physics and Astronomy, UNC-CH. One of us (JHM) ackowledges the receipt of an A. P. Sloan Research Fellowship, an R. J. Reynolds Junior Faculty Development Award, and a grant by the UNC University Research Council.

REFERENCES

1. J. G. Bednorz and K. A. Mueller, Z. *Physik* B**64**, 189 (1986).

2. S. Uchida, H. Tagaki, K. Kitazawa, and S. Tanaka, *Jap. J. Appl. Phys. Lett.* **26**, L1 (1987).

3. C. W. Chu, P. H. Hor, R. L. Meng, L. Gao, Z. J. Huang, and Y. Q. Wang, *Phys. Rev. Lett.* **58**, 405 (1987).

4. R. J. Cava, R. B. Van Dover, B. Batlogg, and E. A. Rietmann, *Phys. Rev. Lett.* **58**, 408 (1987).

5. M. K. Wu, J. R. Ashburn, C. J. Torng, P. H. Hor, R. L. Meng, L. Gao, Z. J. Huang, Y. Q. Wang, and C. W. Chu, *Phys. Rev. Lett.* **58**, 908 (1987).

6. P. H. Hor, L. Gao, R. L. Meng, Z. J. Huang, Y. Q. Wang, F. Forster, J. Vassilious, C. W. Chu, M. K. Wu, J. R. Ashburn, and C. J. Torng, *Phys. Rev. Lett.* **58**, 911 (1987).

7. R. J. Cava, B. Batlogg, R. B. Van Dover, D. W. Murphy, S. Sunshine, T. Siegrist, J. P. Remeika, E. A. Rietmann, S. Zahurak, and G. P. Espinosa, *Phys. Rev.* **58**, 1676 (1987).

8. R. J. Birgeneau, C. Y. Chen, D. R. Gabbe, H. P. Jenssen, M. A. Kastner, C. J. Peters, P. J. Picone, Tineke Thio, T. R. Thurston, H. L. Tuller, J. D. Axe, P. Böni, and G. Shirane, *Phys. Rev. Lett.* **59**, 1329 (1987).

9. Y. Tokura, J. B. Torrance, A. I. Nazzal, T. C. Huang, and C. Ortiz (submitted to J. Amer. Chem. Soc.).

10. J. B. Torrance, Y. Tokura, A. Nazzal, and S. S. P. Parkin (submitted to Phys. Rev. Lett.).

11. G. J. Clark, A. D. Marwick, R. H. Koch, and R. B. Laibowitz, *Appl. Phys. Lett.* **51**, 139 (1987).

12. N. F. Mott, *Conduction in Non-Crystaline Materials*, (Oxford, 1987), p.28.

17

Effects of Surface Properties on Metal-Oxide Superconductor Contact Resistance

YONHUA TZENG Auburn University, Electrical Engineering Department,
200 Broun Hall, Auburn, Alabama 36849

INTRODUCTION

The recent discovery of ternary metal oxide ceramic superconducting materials (Tc 98K) has led to a sudden burst of research activity all over the world (1-9). The demonstration of $10^5 A/cm^2$ critical current density at 77K, (7) the boost of transition temperature to 155K by adding flourine atoms to the superconductor structure (10) and some evidence of room temperature superconductivity further promoted the promise and interest of these materials. Processing of these materials into useful forms such as wires, rods, tubes, films, ribbons, ... etc. will make it possible for applications to high-field magnets, power transmission lines, magnetic coil, microelectronics ... etc.

The applications of the new class of very exciting superconductor materials to microelectronics and microsensors rely on the research to solve the problems of relatively low superconductor critical current density for non-single-crystalline

materials   and   the   poor   metal   to   superconductor   contacts.
Scientists in IBM have demonstrated the   superconducting   critical
current   density   above $10^5 A/cm^2$ for   single   crystalline $YBa_2Cu_3O_6+x$
thin   films   by   means   of   magnetic   measurements.   The   contact
resistance   was   too   high to measure the superconducting critical
current density directly.   Therefore, a more   detailed   study   and
optimization of the metal to superconductor contacts is desirable.
Almost   every   application   of the high temperature superconductor
materials requires a good metal-superconductor contact.

EXPERIMENTAL

Ternary   metal   oxide $YBa_2Cu_3O_{6+x}$ is   fabricated by mixing Y_2O_3,
$BaCO_3$, and CuO in appropriate proportion, burning in   the   air   at
$950^\circ C$ for 6 hrs., and sintering in an oxygen flowing tube at $950^\circ C$
for   12   hrs.   Aluminum   shadow   masks   are   used   to define four
circular dots of   metal deposition on the   surface   of   the
hgih   temperature   superconductor   (HTSC).   Copper is deposited by
the DC sputtering and gold is deposited by a filament evaporator.

        The   deposited   four   dots   are   used   to   do   the   four point
resistance measurement in order to check the superconductivity   of
each   sample.   Two   metal   dots   are   selected   for measuring the
two-point resistance.   In   the   two-point   measurement,   a   desired
current   is   passed   through   two metal dots on the superconductor
using two wires and the voltage across these two dots is   measured
by   another   pair   of   wires   connecting these two metal dots to a
voltmeter. When the substrate   is   superconducting   the   measured
voltage   to   current ratio should be the contact resistance of two
metal-superconductor contacts.

RESULTS AND DISCUSSION

The metal-oxide superconductor contact resistance is very dependent of the surface conditions of the samples. Although some contacts to different batches of samples show a decreasing contact resistance with a decreasing temperature, most of our samples have contact resistance increasing with decreasing temperature. This is opposite to the trend of superconductor resistance. Although the superconductivity critical current density can be high at a low temperature the poorer contact makes the lower temperature not beneficial. More work certainly is needed to improve this contact resistance.

The typical voltage-current characteristics are shown in Figure 1 at three temperatures. These curves show that the

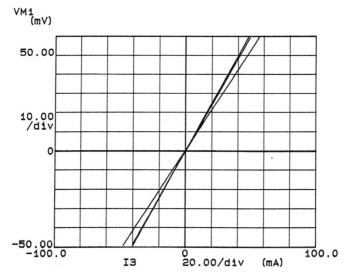

Figure 1.   Typical voltage-current characteristics of a metal-oxide superconductor contact at three temperatures.

contact is ohmic. Four HTSC samples from the same batch with different surface conditions are deposited with copper to form ohmic contacts. The ohmic contact on the fresh HTSC sample shows increasing contact resistance with decreasing temperature. The measured two-point resistance decreases a little when the HTSC sample reaches the superconductivity state and then increases again when the temperature decreases further. Since the two-point resistance is the summation of the contact resistance and the bulk resistance, the slight decrease of two-point resistance is attributed to the decrease of bulk resistance. The specific contact resistance of this copper to fresh sample contact is about 4×10^{-3} ohm-cm^2 at 77°K.

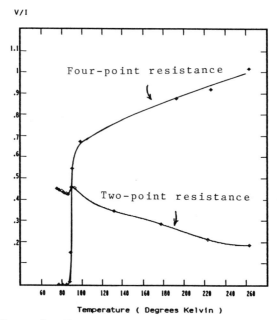

Figure 2.   Two-point and four-point resistance measurements for a fresh HTSC sample with copper contacts. The specific contact resistance at 77°K is 4×10^{-3} ohm-cm^2.

One of these four samples is heated in a vacuum oven at $300^{\circ}C$ for 2 hours before copper is sputtered on it to form ohmic contacts. The vacuum baking accelerates the degradation of the surface layer by reducing the oxygen concentration. This is evidenced by the XPS analysis of the O1S peaks before and after vacuum baking. Copper ohmic contacts to this sample shows rapid increase in contact resistance when temperature decreases. The overall trend of R_c v.s. T is similar to that of the fresh sample. Both the four-point and two-point resistances for this sample is shown in Figure 3. The specific contact resistance is about 2×10^{-2} ohm-cm^2 at $77^{\circ}K$. This is 5 times higher than that of a fresh sample.

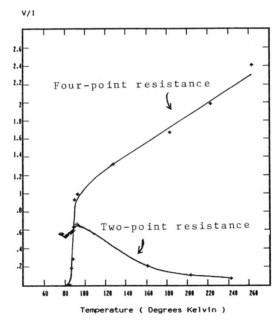

Figure 3. Two-point and four-point resistance measurements for a vacuum baked HTSC sample with copper contacts. The specific contact resistance at $77^{\circ}K$ is 2×10^{-2} ohm-cm^2.

The third sample is baked in air at 500°C for 2 hours before copper contacts are formed. The rate of increase of contact resistance with decreasing temperature is between that of a fresh sample and a vacuum baked sample. The two-point and four-point resistances are shown in Figure 4. The specific contact resistance is 8×10^{-3} ohm-cm^2 at 77°K.

The fourth sample is also a fresh sample. Before the copper deposition, the sample is exposed to an oxygen plasma for 5 minutes without intentional heating. After the oxygen treatment, the sample is deposited with copper to form contacts without exposing to the atmosphere. The two-point resistance decreases

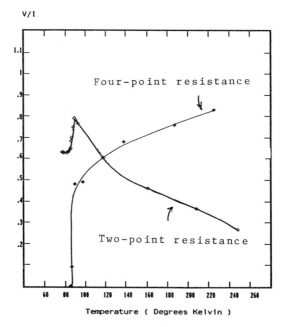

Figure 4.   Two-point and four-point resistance
             measurements for a copper contact to a
             HTSC sample baked in air at 500°C for
             2 hours. The specific contact resistance
             is 8×10^{-3} Ω-cm^2.

with decreasing temperature just like the bulk resistance does. This is shown in Figure 5. The oxygen treatment has apparently recovered at least partially the oxygen deficiency at the surface of the HTSC sample and made the ohmic contacts more desirable. The specific contact resistance is around 1×10^{-3} ohm-cm^2 at $77^\circ K$. Further improvement of the contact resistance requires the compensation of the segregation and pile-up of elements in the ternary metal oxide at the sample surface after high temperature treatments. We are currently working on this subject.

In addition to copper contacts, gold contacts are also studied. The gold contact to a fresh sample also shows increasing contact resistance with decreasing temperature as shown in Figure

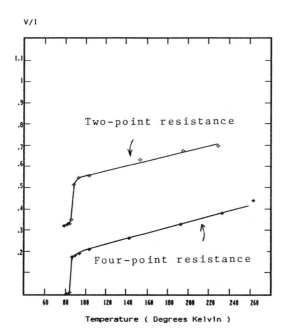

Figure 5.   Two-point and four-point resistance measurements for a copper contact to a HTSC sample treated with an oxygen plasma. The specific contact resistance is 1×10^{-3} ohm-cm^2.

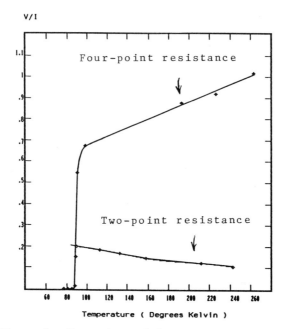

V/I

Figure 6.   Two-point and four-point resistance
             measurements for a gold contact to a
             fresh sample.   The specific contact
             resistance is 8 x 10^{-3} ohm-cm2 at 77°K.

6.   The   specific contact resistance is 8 x 10^{-3} ohm-cm^2 at 77°K.
After the gold contact is heated to 450°C in the air for 1   hour,
the gold apparently diffuses into the HTSC as the color of the
gold contacts turns into dark brown.   This   causes the   contact
resistance   to   increase   significantly as shown in Figure 7.   The
specific contact resistance increases from 8 x 10^{-3} ohm-cm^2 to 6 x
10^{-1} ohm-cm^2 at 77°K.

CONCLUSION

Ohmic   contacts   between a metal layer and an oxide superconductor
prepared by a simple deposition of metal onto   the   oxide

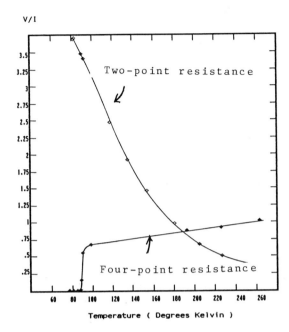

Figure 7.   Two-point and four-point resistance
measurements for a gold contact to a
fresh sample followed by a heat treatment at 450°C for 1 hour. The specific
contact resistance is 6 x 10^{-1} ohm-cm^2
at 77°K.

superconductor surface have poor contact resistance.   Oxygen

plasma treatment of the sample surface before metal deposition

improves the contact resistance. A better and maybe more

sophisticated metal-oxide superconductor interface is required to

achieve acceptable ohmic contacts.

ACKNOWLEDGEMENTS

This work was partially supported by the Alabama Microelectronics

Science and Technology Center and the Space Power Institute. Very

valuable technical assistance provided by Professor M.K. Wu is

appreciated.

REFERENCES

1. M.K. Wu, J.R. Ashburn, C.J. Torng, P. H. Hor, R. L. Meng, L. Gao, Z.J. Huang, Y.O. Wang and C. W. Chu, Phys. Rev. Lett., 58, 908 (1987).

2. C.W. Chu, P.H. Hor, R.L. Meng, L. Gao, Z. J. Huang and Y.O. Wang, Phys. Rev. Let. 58, 405 (1987).

3. J.G. Bednorz and K. A. Muller, Z. Phys. B 64, 189 (1986).

4. J. G. Bednorz, M. Takashige and K. A. Muller, Europhys. Lett., 3, 379 (1987).

5. Xhang Yuling, Xie Sishen, Cheng Xiangrong, Yang Qianshong, Liu Guirong and Ni Yongming, J. Phys D. Appl. Phys. 20, 14, (1987).

6. S.K. Makik, A. M. Umarji, D.T. Adroja, C. V. Tomy, Ram Prasad, M.C. Soni, Ashok Mohan, and C.K. Gupta, J. Phys. C: Solid State Phys. 20, L347. (1987).

7. P. Chandhari, R. H. Koch, R. B. Laibowitz, T. R. McGuire, and R. J. Gambino, Submitted.

8. S.K. Dhar, P.L. Paulose, A.K. Grover, E. V. 17, L105, (1987).

9. P.L. Paulose, V. Nagarajan, A.K. Gover, S.K. Dhar and E.V. Sampathkumaran, J. Phys. F: Met. Phys. 17, L91, (1987).

10. S.R. Ovshinsky, R.T. Young, D. D. Allred, G. DeMaggio, and G.A. Van der Leeden, Phys. Rev. Lett., 58, 2579, (1987).

18

Observations of the Deterioration of YBa$_2$Cu$_3$O$_7$ Using a New Superconductor Characterization Cryostat

RALPH C. LONGSWORTH and WILLIAM A. STEYERT   APD Cryogenics Inc, Allentown, Pennsylvania

1. INTRODUCTION

A Comparatively simple test unit has been developed which is referred to as a Superconductor Characterization Cryostat (SCC). It is suitable for a wide variety of measurements, requires no cryogenic refrigerants, and requires very little cryogenic expertise to use (1).

Cooling for the SCC is provided by a Displex® CS202 two stage low temperature refrigerator capable of cooling below 10 K. The cold end of the CS202 is thermally attached to a 19 mm ID beryllium copper sample tube (aluminum is available for x-ray work) which can be cooled to 12 K by the refrigerator. Samples of up to 12.7 mm in diameter and 150 mm long can be inserted into the sample area. A heat exchange gas introduced into the sample area cools the sample to the temperature of the beryllium copper tube. The sample area can be evacuated for specific heat work. A simple air lock mechanism at the warm end of the 19 mm sample space allows for fast sample change without introducing gaseous contaminants into the sample area and without appreciable warming of the refrigerator and sample area. A heater and thermometer on the cold end of the

---

®
Trademark of APD Cryogenics

refrigerator allow regulation of the sample area temperature from 12
K to over 350 K with a stability of a small fraction of 1 K. The
temperature at the sample can also be measured with a thermocouple
or diode temperature sensor. The SCC is shown in Figure 1.

This unit has been used to study the electrical resistivity
$YBa_2Cu_3O_7$ (1-2-3). Experiments have shown that after being
cooled and exposed to moisture, the 90 K superconductivity can be
lost.

2. USES OF THE SCC

2.1 Four Terminal Resistance

The sample is attached to the end of the holder. Two leads are
attached which carry low frequency alternating current, direct
current, or pulsed currents through the sample. Two separate
leads pick up the voltage developed by the passage of this
current through the sample. When the sample is cold and
superconducting, this voltage is zero. As the sample warms,
either under control of the refrigerator temperature controller
or by turning the refrigerator off and allowing a drift, at some
temperature, there is a voltage developed. As the temperature
increases further, the resistance increases sharply until the
superconducting-to-normal transition is complete. Depending on
one's definition, T_c is at the beginning, middle, or upper
temperature of this transition. Sensitivities of better than
10^{-6} V are possible with DC and about 10^{-9} V with AC. The
SCC is supplied with four terminal sample mounting clips.

A solenoid or permanent magnet (32 mm ID) located in the air
outside the sample area allows study of the dependence of T_c on
small fields.

2.2 Other Uses of the SCC

The SCC is designed to make a wide variety of measurements on the
new superconductors:

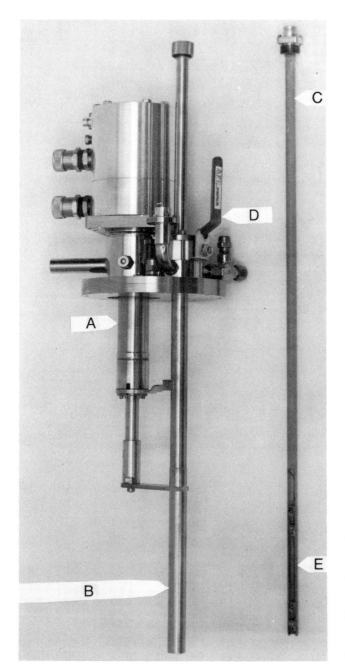

A. REFRIGERATOR
B. SAMPLE WELL
   12 K to 350 K
C. SAMPLE HOLDER
   WITH 12-PIN
   FEEDTHROUGH
D. AIR LOCK FOR
   FAST SAMPLE
   CHANGE
E. SAMPLE

Figure 1.   Superconductor Characterization Cryostat, SCC, with vacuum shroud removed.   G-10 sample holder inserts into sample well and is surrounded with helium exchange gas.

- Four Terminal Critical Current
- Hall Voltage (Charge Carrier Density)
- Specific Heat
- X-Ray Diffraction
- Dielectric Constant
- Thermal Conductivity
- Tunneling
- AC Susceptibility (Meissner Effect)
- Leadless J_c (2)
- Thermoelectric Coefficient

3. DC RESISTANCE MEASUREMENT OF 1-2-3 SUPERCONDUCTOR

A fresh sample of 1-2-3 was shaped by grinding into the form of a
crude cylinder 2-1/2 mm in diameter and 7 mm long. Two leads were
attached separately with silver micropaint at each end of the
cylinder. Current (0.1A) was run through two leads, and the
voltage measured on the other leads. Each current-on measurement
was accompanied by a current-off measurement (to correct for any
thermoelectric effects). The current density is 2 A/cm^2; this
is probably the cause of the 85 K transition temperature, rather
than the expected 90 K transition temperature. Figure 2 shows the
experimental results.

In Figure 2a, freshly mounted sample, the resistivity below
the transition is 0 ±1 μΩ cm. However, for the same sample
measured the next day, Figure 2b, with the same experimental
setup, the resistivity just below the transitions measured 8.5
±1 μΩ cm. Below 37 K, the resistivity returned to 0 ±2 μΩ
cm.

4. SUMMARY

A new cryostat has been described which is a very versatile tool
for characterizing new high temperature superconducting
materials. It has been used to measure the DC resistivity of the

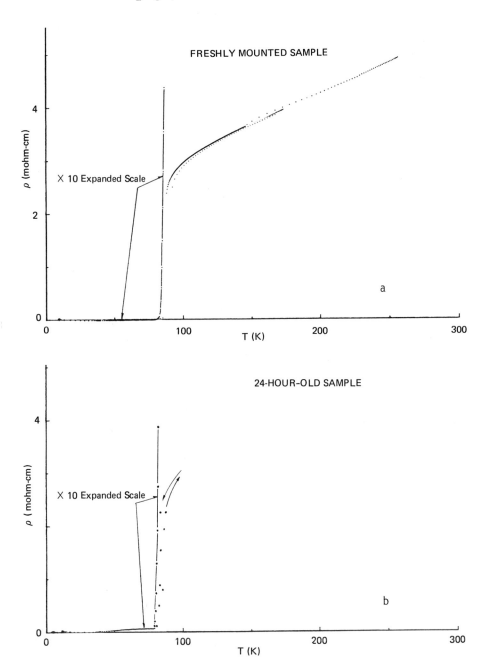

Figure 2. Four terminal resistance; data system has reduced measured voltages to resistivities and plotted ρ vs. T. This data, and similar plots not presented, represent many temperature cycles of consistent results.

new 1-2-3 material.   Results show significant deterioration after
a 24 hour exposure to atmospheric air.

REFERENCES

1.   SCC is available from APD Cryogenics Inc, Allentown, Pa.

2.   C. P. Bean, Rev. Mod. Phys., 36, 31 (1964).

19

Superconductivity, Thermogravimetry, EPR, Electron Microscopy, and X-Ray Diffraction of $YBa_2Cu_3O_{7-z}$

ANNY MORROBEL-SOSA, DAVID A. ROBINSON, CHINNARONG ASAVAROENGCHAI, ROBERT M. METZGER, JOSEPH S. THRASHER Department of Chemistry, The University of Alabama, Tuscaloosa AL 35487-9671

CHESTER ALEXANDER, Jr. Department of Physics, The University of Alabama, Tuscaloosa AL 35487-1921

DONALD A. STANLEY and M. ABBOT MAGINNIS U.S. Bureau of Mines, Tuscaloosa Research Center, Tuscaloosa AL 35486-9777

1. INTRODUCTION

The study of superconductivity has undergone various transformations since the initial discovery of the phenomenom by Kamerlingh Onnes(1); yet, none have generated the enormous flurry in research activity in the field as have the more recent discoveries of Bednorz and Mueller (2) and that of Chu, Wu and colleagues (3). Reports of high-temperature superconductivity in rare-earth copper oxides now come from many laboratories worldwide, including our own (4-6).

These new materials consist of a combination, in varying molar proportions, of R-A-Cu-O, where R is a lanthanide rare-earth ion and A is an alkaline earth ion (usually barium or strontium). It is now clear that there exist at least two families of these metallic oxides that exhibit superconductivity: the $La_{2-x}A_xCuO_{4-z}$ family, with T_c's in the 30-50 K range, and with a tetragonal layered perovskite-type (K_2NiF_4) structure, and the $RBa_2Cu_3O_{7-z}$ family, with T_c's in the 90-100 K range, and with an orthorhombic distorted perovskite structure. In both families there are planar or nearly planar arrays of Cu and O ions of stoichiometry CuO_2 There is also a ribbon motif of Cu and O atoms in the $YBa_2Cu_3O_{7-z}$

structure of stoichiometry CuO_3, which may be responsible for the higher T_c (7).

Recently we reported on superconductivity in a green-black composite $Gd_3Ba_3Cu_4O_z$, which, when quenched within seconds from 1170 K to 77 K, was found to superconduct with T_c = 62 K (onset of superconductivity at 71 K) (6). The report presented here concerns our most recent efforts in the study of the superconductivity in single-phase $YBa_2Cu_3O_{7-z}$ (8).

2. EXPERIMENTAL

The studies of $YBa_2Cu_3O_{7-x}$ reported here were prepared from fine powered samples of Y_2O_3 (Molycorp 99.99%), $BaCO_3$ (Mallinckrodt) and CuO (Aldrich) in the mole ratio 0.5:2:3. The materials were sintered at 950°C in a Lindberg tube furnace under flowing oxygen for 24 h. Pressed pellets (1.2 kbar) were made after the sample was furnace cooled to room temperature. These pellets were then sintered at 950 °C for 8 h, annealed at 500°C for 6 h, and cooled slowly to 200°C, all done in the presence of flowing oxygen. If preliminary dc electrical resistance measurements did not indicate the presence of a superconducting transition, the materials were sintered further for 12 h at 900°C, and for 5 h at 500°C.

Measurements of dc electrical resistance versus temperature were performed on samples attached to, and electrically insulated from, the cold finger of a closed cycle refrigerator (CTI-Cryogenics 22C). Four parallel leads were painted onto the samples with conductive silver paint (DuPont 7713), and the small amount of solvent is removed by evaporation at 540°C for about 10 minutes, in flowing oxygen. Temperature was measured with a chromel versus Au-0.07% Fe thermocouple also attached to the sample with the same electrically insulating varnish (GE7031). Copper wire contacts were soldered directly onto the leads. The measurements were made by reading the voltage drop across the inner leads of the sample with a nanovolt amplifier (Keithley 140) while maintaining a constant sample current of less than 1.5 mA. The temperature data were digitized with a digital temperature indicator (Scientific Instruments). Data acquisition and analysis were performed with a personal computer (HP-9816) and a data acquisition/control unit (HP-3497A).

Thermogravimetric analysis (DuPont 1090) was performed in argon on a 60 mg sample. The EPR spectra were studied at 9.5 GHz using a Varian E-12 spectrometer equipped with a frequency counter (HP-5246L) and NMR gaussmeter (Bruker ER-035M). Spectra at 77 K were obtained using a Varian variable temperature dewar. The spectrometer was calibrated with a standard ruby source (NBS). X-ray diffraction (Phillips 3100), with Cu K_α radiation, SEM (ATEC Autoscan and AMR-1000), and chemical analysis (Galbraith Laboratories) were also performed on these samples.

3. RESULTS AND DISCUSSION

The chemical analysis for $Y_1Ba_2Cu_3O_{6.85}$ was: Calc (obs): Y: 13.39% (14.01), Ba: 41.38% (40.44), Cu:28.72% (28.39). SEM revealed the polycrystalline nature of the samples with typical crystallite size of 60 μ. TGA revealed a 0.61 mass% loss around 275°C, a loss of 1.28% centered in

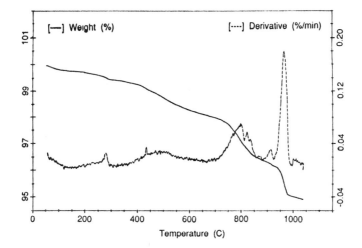

Fig. 1. Thermogravimetric analysis of mass loss (_____) and of its first derivative (- - - -) for YBa2Cu3O7-z as a function of temperature.

a broad 150°C range about 500°C, a 1.81% loss around 800°C, and a 1.43% loss about 985°C. It is probable that the mass losses are associated with loss of O_2 from the lattice (1 O_2 per $YBa_2Cu_3O_{6.85}$ = 4.82 mass%). The X-ray diffraction pattern consists of 35 lines, of which only 5 very weak lines could not be indexed for the known structure (9), although the intensities match closely.

The results for the four-probe dc resistance versus temperature measurements are shown in Fig. 2. The onset temperature for superconductivity was 102 K and the temperature where zero resistance was reached was 96K. This result is consistent with the T_c values reported by other groups (10,11).

The EPR spectrum is given in Fig. 3. There is one weak, broad, anisotropic powder line ($g_{||}$ = 2.128, g_{perp} = 2.067, linewidth 210 Gauss). If the formula is assumed to be $Y_1Ba_2Cu_3O_{6.85}$, then one expects 0.726 Cu^{++} per "molecule"; however, the measured intensity is only 0.85% of the expected intensity. The discrepancy between the measured and

Table 1

X-ray powder diffractogram for $Y_1Ba_2Cu_3O_{7-z}$ [λ (Cu Kα_1) = 1.540598 Å, λ(Cu Kα_{ave}) = 1.541877 Å], was indexed to yield a=3.824(1) Å, b=3.891(1) Å, c=11.690(3) Å.

h k l	d/Å	I/I$_o$	h k l	d/Å	I/I$_o$	h k l	d/Å	I/I$_o$
0 0 1	11.570	3	0 1 3	2.750	22	2 0 3	1.716	1
0 0 2	5.818	11	1 0 3	2.729	30	0 0 7	1.670	13
0 0 3	3.890	43	1 1 1	2.651	1		1.668	12
0 1 0		——		2.519	<1	1 1 6	1.586	13
1 0 0	3.812	1	1 1 2	2.469	1	2 1 3	1.570	6
0 1 2	3.236	1	0 0 5	2.336	73	0 1 7	1.533	1
1 0 2	3.195	2	1 1 3	2.232	6	0 2 5	1.496	1
——	3.147	1	1 1 4	1.990	1	2 0 5	1.480	1
——	3.064	1	0 0 6	1.948	100	0 0 8	1.460	1
——	2.990	1	2 0 0	1.911	6	1 1 7	1.425	1
0 0 4	2.915	5	1 1 5	1.774	2	0 2 6	1.377	3
——	2.827	1	0 2 3	1.738	1	2 2 0	1.365	8

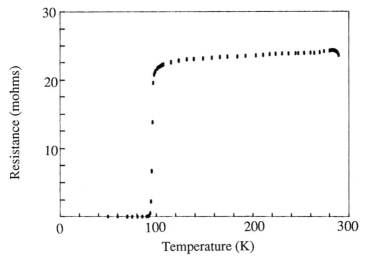

Fig.2. Resistance of a pellet of YBa2Cu3O7-z as a function of temperature.

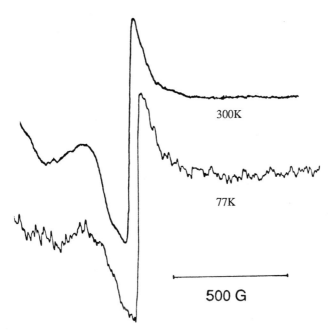

Fig. 3. Electron paramagnetic resonance of YBa2Cu3O7-z (at 300 K and 77K).

expected intensity may be processing dependent, such as that induced by impurities (10). Yet, the observed spectra may also be attributable to antiferromagnetic coupling between Cu^{++} ions. Our results are consistent with a recent report on the temperature dependence of the EPR for $YBa_2Cu_3O_{7-z}$ (11).

ACKNOWLEDGMENTS

We are grateful to Prof. G. N. Coleman of the Department of Chemistry for providing many of the starting materials, Prof. J. H. Fang of the Department of Geology for providing the X-ray data, and to Dr. Deborah L. Clayton of the Biology Department for the scanning electron microscopy. One of us (D.A.R.) would also like to acknowledge the U.S. Bureau of Mines for the granting of a Research Fellowship for the projects reported here.

REFERENCES

1. H. Kamerlingh Onnes, Akad. van Wetenschappen (Amsterdam) **14,** 113, 818 (1911).

2 J. G. Bednorz and K. A. Mueller, Z. Phys. B **64**, 189 (1986).

3. M. K. Wu, J. R. Ashburn, C. J. Torng, P. H. Hor, R. L. Meng, L. Gao, Z. J. Huang, Y. Q. Wang, and C. W. Chu, Phys. Rev. Lett. **58**, 908 (1987).

4. D. U. Gubser and M. Schutler, eds., Extended Abstracts - High-Temperature Superconductors, Session S, Materials Research Society , (Anaheim CA 23-24 April 1987), and references therein.

5. D. L. Nelson, M. S. Whittingham, and T. F. George, eds., Chemistry of High Temperature Superconductors, A.C.S. Symposium Series **351** (New Orleans LA, Aug. 30 - Sep. 4, 1987), and references therein.

6. D. A. Robinson, A. Morrobel-Sosa, C. Alexander, Jr., J. S. Thrasher, C. Asavaroengchai, and R.M. Metzger, in Proceedings of the NATO Workshop on Inorganic and Organic Low-Dimensional Crystalline Materials (Minorca, 3-8 May 1987), P. Delhaes, Ed., Plenum, in press.

7. M. H. Whangbo, M. Evain, M. A. Beno, and J. M. Williams, Inorg. Chem., **26**, 1831 (1987).

8. D. A. Robinson, A. del C. Morrobel-Sosa, C. Alexander, Jr., J. S. Thrasher, C. Asavaroengchai, R.M. Metzger, D. A. Stanley, and M. A. Maginnis, J. Chem. Phys., submitted.

9. R. J. Cava, B. Batlogg, R. B. van Dover, D. W. Murphy, S. Sunshine, T. Siegrist, J. P. Remeika, E. A. Rietman, S. Zahurak, and G. P. Espinosa, Phys. Rev. Letters 58, 1676 (1987).

10. G. J. Bowden, P. R. Elliston, K. T. Wan, S. X. Dou, K. E. Easterling, A. Bourdillon, C. C. Sorrell, B. A. Cornell, and F. Separovic, J. Phys. C, in press.

11. R. Jones, M. F. Ashby, A. M. Campbell, P. P. Edwards, M. R. Harrison, A. D. Hibbs, D. A. Jefferson, A. I. Kirkland, T. Thanyasiri, and E. Sinn, in Chemistry of High-Temperature Superconductors, D. L. Nelson, M. S. Whittingham, and T. F. George, eds., A.C.S. Symposium Series 351 (New Orleans LA, Aug. 30-Sep. 4, 1987), p. 313.

20

Optical Properties of La$_{2-x}$Sr$_x$CuO$_4$

S.L. HERR, K. KAMARÁS, C.D. PORTER, M.G. DOSS and D.B. TANNER
Department of Physics, University of Florida, Gainesville, Florida

D.A. BONN, J.E. GREEDAN, C.V. STAGER and T. TIMUSK Institute for
Materials Research, McMaster University, Hamilton, Ontario, Canada

S. ETEMAD, D.E. ASPNES, M.K. KELLY, R. THOMPSON, J.-M. TARASCON
and G.W. HULL Bell Communications Research, Red Bank, New Jersey

I.   Introduction

The optical properties of the high T_c superconductor La$_{2-x}$Sr$_x$
CuO$_4$ have been investigated as a function of temperature (T=4 to
300K) and Sr concentration (0 $\leq$ x $\leq$ 0.30) from the far infrared to
the ultraviolet spectral region. Kramers-Kronig analysis of the
reflectance shows a frequency dependent conductivity with three
strong vibrational features (240, 355 and 495 cm^{-1}) in the far
infrared and two electronic features (3,500 and 10,500 cm^{-1}) in
the infrared region. The large oscillator strength of the 240 cm^{-1}
vibrational feature rules out an interpretation in terms of an
ordinary optical phonon, whereas the energy of the low-lying electronic transition is inconsistent with a single-particle picture
of the electronic band structure. The concentration dependence of
the oscillator strengths for both vibrational and electronic
features are correlated with T_c as well as with the fractional
ideal diamagnetism (Meissner effect) indicating that there is a
connection between these properties and superconductivity (1).

II. Experimental Details

  A. Temperature Dependence of x=0.15 Sample

    Samples of $La_{1.85}Sr_{0.15}CuO_4$ were prepared by previously
described ceramic techniques (2,3) to form sintered, pressed pellets. Samples for this study were single phase (x=0.15) with the
midpoint of a 1.6K wide superconducting transition at 40.3K, but
an incomplete Meissner effect. The optical measurements were made
as near-normal-incidence reflection on polished sample surfaces.
Measurements in the far infrared were done at McMaster University
while the mid-infrared was done at the University of Florida, as
described elsewhere (2). Following the measurements the samples
were coated with a metal layer (Pb or Al) in order to provide a
reference reflectance. On account of the granular nature of the
surface this coating made important quantitative changes in the
absolute value of the reflectance, particularly at higher
frequencies.

  B. Concentration Dependence

    The series of samples $La_{2-x}Sr_xCuO_4$ are semiconducting for
$x \leq 0.01$, superconducting for $0.03 < x < 0.30$ and metallic for
$x \geq 0.30$ (1). Room temperature optical data for the far infrared
to mid infrared were measured as near-normal incidence reflection
as discussed above higher frequency measurements were made at Bell
Communications Research using spectroreflectometry for the 2,400
to 24,000 cm^{-1} region and spectroellipsometry for the 12,000 to
46,000 cm^{-1} region. The experimental details are described elsewhere (1,4). Reflectivity measurements at Bellcore were referenced
to gold coated samples for an absolute value of reflectance. The
sample surfaces were polished to near mirror quality so that the
reflectance was scaled to the Bellcore ellipsometry data.

III. Results

  A. Temperature Dependence

    Figure 1 shows the reflectance at 4K and 300K. Except
for the frequencies below 200 cm^{-1}, where the effects of

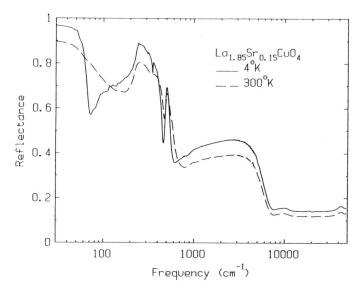

Figure 1.   Reflectance of La$_{1.85}$Sr$_{0.15}$CuO$_4$ at 4 and 300K.   Note
the logarithmic frequency scale.

superconductivity are seen (2), the data at 70K are extremely
close to those at 4K. Warming to 300K gives a decrease in the
mid-infrared reflectance.

The frequency dependent conductivity, obtained from Kramers-
Kronig analysis of the reflectance, is shown in Figure 2.   The
electronic part of the spectrum is decidedly non-Drude-like with a
broad peak at 3,500 cm^{-1} and a weaker maximum at 10,500 cm^{-1}.   The
3,500 cm^{-1} feature is more significant than it appears in Figure
2; the logarithmic scale compresses it. At low frequencies, the
conductivity is dominated by a vibrational feature at 240 cm^{-1};
weaker peaks are seen at 355 and 495 cm^{-1}. None of the peaks displays a particularly strong temperature dependence. Finally, we
note that the low-frequency conductivity, which at 300K is about
200 Ω^{-1}cm^{-1}, is a little larger than the normal state dc conductivity (~40 Ω^{-1}cm^{-1}), a discrepency that may be due to the fact that
this is a polycrystalline sample of anisotropic material.

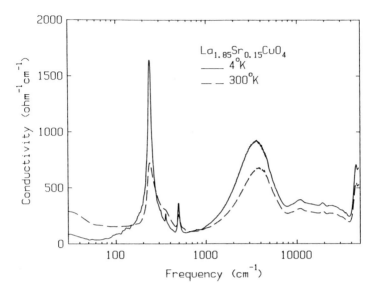

Figure 2.   The frequency dependent conductivity of
            La$_{1.85}$Sr$_{0.15}$CuO$_4$, at 4 and 300K, as determined by
            Kramers-Kronig analysis of the reflectance.

B.   Concentration Dependence

        The reflection spectra of a series of La$_{2-x}$Sr$_x$CuO$_4$
samples are shown in Figure 3. The logarithmic frequency scale
emphasizes the low frequency region as the higher frequencies
(above 7,000 cm^{-1}) show little concentration dependence. (See
reference 1 for more detail of this region.) In the region 200-
7,000 cm^{-1} the reflectance features increase non-linearly with x
reaching a maximum near x=0.15 and then decrease. We note the
onset of the peak at 240 cm^{-1} at low concentration (x=0.03), which
persists to the highest Sr concentration. Also notable is the
disappearance of the small feature 650 cm^{-1} for x=0.10 and its re-
emergence at x=0.30. The result of primary importance is the dom-
inance of the electronic feature between 600 and 7,000 cm^{-1}, which

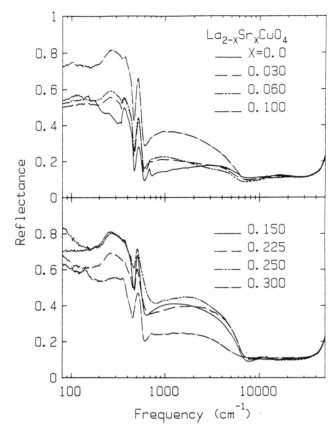

Figure 3. Room temperature reflectance of a series of
           La$_{2-x}$Sr$_x$CuO$_4$ concentration.

shows a dramatic concentration dependence. The frequency dependent
conductivity, Figure 4, clearly shows that the maximum oscillator
strength for this electronic feature as well as the vibrational
features at 240 and 495 cm^{-1}, is near x=0.15-0.20.

IV. Discussion

The oscillator strengths of the vibrational (240 cm^{-1}) and
the electronic (3,500 cm^{-1}) peaks are the key to interpreting

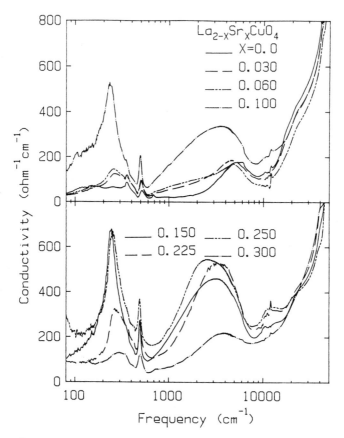

Figure 4. The frequency dependent conductivity determined from the reflectance for a series of $La_{2-x}Sr_xCuO_4$ concentrations, obtained by Kramers-Kronig analysis of the reflectance.

these measurements. The strengths of the vibrational feature is too intense for it to be an ordinary optical phonon, whereas the electronic peak has too little oscillator strength to be attributed to nearly free carriers. The occurrence of a strong electronic band in the infrared spectrum and its correlation with superconductivity (5) provide evidence that the superconductivity of these materials may be mediated by electronic excitations.

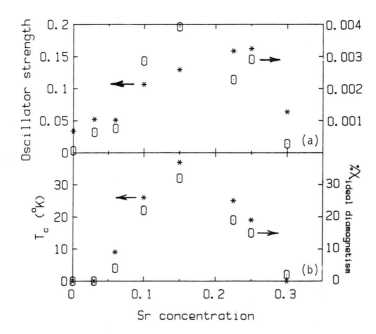

Figure 5.   a) Oscillator strength of the vibrational mode at 240
             cm^{-1} (θ) and the electronic absorption band at 3,500
             cm^{-1} (*), as a function of Sr concentration; b)
             Fractional ideal diamagnetism (Meissner effect) (O) and
             T$_c$ (*) as a function of Sr concentration.

The oscillator strengths of the vibrational (240 cm^{-1}) and
electronic modes as a function of dopant concentration are shown
in Figure 5a. Below this, Figure 5b, the fractional ideal diamagnetism (Meissner effect) and Tc are plotted versus concentration.
The increased oscillator strength of the electronic absorption
feature follows both T$_c$ and the fractional ideal diamagnetism as
might be expected for a superconductivity mediated by an electronic excitation. It is evident that the oscillator strength of the
vibrational band increases in proportion to that of the electronic
absorption. The oscillator strengths of the dopant-induced bands
correlate with the superconducting properties as the concentration
is changed to cover the full range corresponding to the appearance
and disappearance of superconductivity.

In summary, the optical properties of $La_{2-x}Sr_xCuO_4$ are consistent neither with a free-electron picture for the conduction electrons nor with the vibrational feature being a simple optical phonon. We have shown a positive correlation between superconductivity and intense dopant induced bands in this material. The data shows that the electrons are strongly interacting - among themselves and with the vibrational modes.

ACKNOWLEDGMENTS

Research at the University of Florida was supported by the National Science Foundation Solid State Chemistry Grant No. DMR-8416511, and by the Defense Advanced Research Projects Administration through a grant monitored by the Office of Naval Research. Research at McMaster University was supported by the Natural Sciences and Engineering Council of Canada.

REFERENCES

1.   S. Etemad, D.E. Aspnes, M.K. Kelly, R. Thompson, J.-M. Tarascon and G.W. Hull, preprint.

2.   D.A. Bonn, J.E. Greedan, C.V. Stager, T. Timusk, M.G. Doss, S.L. Herr, K. Kamarás, C.D. Porter, D.B. Tanner, J.-M. Tarascon, W.R. McKinnon and L.H. Greene, Phys. Rev. B35, 8843 (1987).

3.   J.-M. Tarascon, L.H. Greene, W.R. McKinnon, G.W. Hull and T.H. Geballe, Science 235, 1373 (1987).

4.   D.E. Aspnes and A.A. Studna, Appl. Opt. 14, 220 (1975); Rev. Sci. Instrum. 43, 291 (1978).

5.   We have previously shown that the electronic peak disappears for superconducting $YBa_2Cu_3O_{6.9}$ upon removal of oxygen (by heating under vacuum) to form non-superconducting $YBa_2Cu_3O_{6.2}$ indicating that the peak is indeed related to superconductivity in the YBa compound. We believe that the origin of superconductivity is similar for both the YBa and LaSr copper oxides. K. Kamarás, C.D. Porter, M.G. Doss, S.L. Herr, D.B. Tanner, D.A. Bonn, J.E. Greeden, A.H. O'Reilly, C.V. Stager and T. Timusk, Phys. Rev. Lett. 59, 919 (1987).

21

Infrared Reflectance of Rare Earth-Barium-Copper Oxide Superconductors

R. SUDHARSANAN, S. PERKOWITZ, B. LOU, B. R. CALDWELL, and G. L. CARR
Physics Department, Emory University, Atlanta, Georgia 30332

Introduction

Despite considerable efforts, the physical mechanism which gives rise to the remarkably high transition temperatures in the rare earth-barium-copper oxide (RBCO) superconductors is still not understood. Infrared studies[1,2,3,4] of the bulk material, mostly with R=Y, reveal a complicated spectrum rich with phonon structure. One consequence is that the spectroscopic energy gap frequency $\omega_g = 2\Delta/\hbar$ is still not well established. The first reports gave $\omega_g \approx 200-250$ cm^{-1}, so that $\Delta \approx 14$ meV, considerably less than the values determined by tunneling[4,5,6]. Improvements in sample fabrication have yielded material for which the free carriers play a more dominant role in the infrared spectra, including materials with different rare earths.

In this paper we present the results of infrared reflectance measurements on three different high T_c superconducting materials, YBa$_2$Cu$_3$O$_7$, GdBa$_2$Cu$_3$O$_7$, and TmBa$_2$Cu$_3$O$_7$. We discuss the phonon modes, the occurrence of coupled plasmon-phonon modes, and the spectroscopic energy gap. We conclude that the energy gap is probably much higher than the initial estimates of 200-250 cm^{-1}.

Preparation and Characterization

Samples of YBa$_2$Cu$_3$O$_7$, GdBa$_2$Cu$_3$O$_7$, and TmBa$_2$Cu$_3$O$_7$ were prepared from Ba$_2$CO$_3$, CuO, and R$_2$O$_3$ (where R=Y, Gd, or Tm). Appropriate quantities of each powder were thoroughly mixed together using a liquid nitrogen freezer mill, then dried and calcined at 900 °C for 5 hours. The resulting black powders were ground and mixed in the mill, then pressed at 8 kbar to produce 13 mm diameter wafers

from 1 to 2 mm thick. The wafers were baked overnight at
900 °C in a flowing O_2 atmosphere, after which the oven was cooled
over a 5 hour period to below 400 °C before removing the sample.
To enhance uniformity, each wafer was reground, repressed at a
higher pressure of 10 kbar, and baked under oxygen a second time.
The resulting wafers had densities approximately 80% of the ideal
values and smooth surfaces suitable for reflectance measurements
(polishing improved their luster but did not have an appreciable
affect on the infrared reflectance). The resistivities, as
determined using the four probe Van der Pauw technique, were
between 1000 and 1500 $\mu\Omega$-cm at room temperature and a factor of
2 to 3 lower at 100 K. The transitions to the superconducting
state were about 6K wide and centered at 93K for the Y and Gd
samples, and at 87K for the Tm sample.

Experimental Results

The infrared reflectance of each sample was measured at near normal
incidence using a Michelson interferometer with Golay cell detector
set up to cover the range from 25 to 650 cm^{-1} at a resolution of
8 cm^{-1}. Sample cooling was accomplished with a continuous flow of
liquid helium to the sample mount (Air Products Heli-tran). Temperatures of 6K and 110K were used for the superconducting and normal
states respectively. The reflectance from a polished coin silver
mirror was used for reference.

Figure 1 shows the normal state reflectances for the three
sample materials. Numerous phonons appear as peaks in the reflectances with principal modes near the frequencies of 165, 290, 330,
400, and 600 cm^{-1}. The phonon modes for the Y and Gd
samples occur at essentially the same frequencies. However, the
phonons for the Tm sample are somewhat shifted relative to the
modes for either the Y or Gd samples.

In Figure 2 we show the ratio of the reflectance in the superconducting state to that in the normal state. A region of enhanced
reflectance in the superconducting state extends up to about
350 cm^{-1}. At frequencies just below those of the stronger phonons
(i.e 160 cm^{-1}, 250-320 cm^{-1}) the ratio drops to near unity.

Analysis and Discussion

We discuss three features of the infrared reflectance results,
beginning with the phonon spectrum (including the frequency shifts
for the Tm sample), followed by the low plasma frequencies and the
possibility of coupled plasmon-phonon modes, and finally we look
into the question of the energy gap frequency.

The phonon spectrum for Tm, which shows the most deviation
from the other sample materials (Y or Gd), is somewhat surprising
since yttrium is the lightest of the three rare earths studied

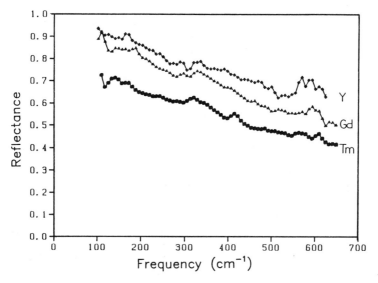

FIGURE 1.   Infrared reflectance of $RBa_2Cu_3O_7$ where R = Y, Gd, and Tm. The curve for R = Tm has been offset by -0.1 for clarity.

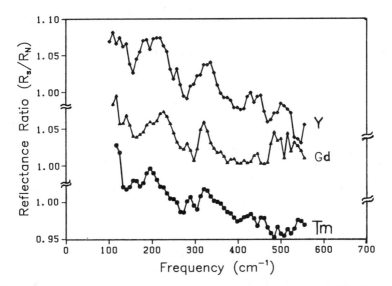

FIGURE 2.   Ratio of the superconducting state reflectance (R_S) to the normal state reflectance (R_N) for $RBa_2Cu_3O_7$ where R = Y, Gd, and Tm.   The curves have been vertically displaced for clarity.

and would be expected to show the greatest deviations from the others. That this does not occur suggests that the deviations are due to a structural difference. Y, Gd, and Tm each form solids in the hcp crystal structure. The lattice parameters[7] for Y and Gd differ from each other by less than 1% whereas Tm differs from Y or Gd by about 3%. A resulting structural distortion around the Tm atoms combined with anharmonic effects could cause mode frequencies to shift up or down, depending on bond stretch or compression. This may be indirectly responsible for the somewhat lower T_c observed for this material.

We have fitted the reflectance data using a dielectric function comprised of a Drude term (to model the free carriers) and a sum of Lorentzian oscillators (to model the phonons). The fits yield values for the bare plasma frequency ranging from 1500 to 4500 cm^{-1}, i.e. the middle infrared. The bare plasma frequency can be used to determine a carrier concentration if one assumes an effective charge and mass equal to those for a free electron. Such a procedure yields values for the concentration in the range of 10^{20} cm^{-3}.

A potentially important consequence of the low plasma frequencies is the existence of coupled plasmon-phonon modes in the middle to far infrared regions[8]. When a zero occurs in the dielectric function, due to the sum of a (negative) free carrier term and a (positive) phonon term, a longitudinal mode having a coupled character occurs. Unless the coupling is weak, these modes appear as reflectance minima in the vicinity of an optical phonon. The Y and Gd reflectances (see Fig. 1) show minima near 160, 280, and 310 cm^{-1}; frequencies just below some of the stronger phonon modes. The degree of coupling depends on damping, but also on the proximity to the phonon mode. Perkowitz[9] has shown that, under certain circumstances, these modes are undamped in the superconducting state and coupling is more likely. Such modes can be important for superconductivity. In particular, Cohen[10] has discussed how low carrier densities affect superconductivity and notes that coupled plasmon-phonon modes can enhance T_c in some instances.

The problem of determining the energy gap is difficult. In a conventional metallic superconductor, such as tin, the absorption edge appears as a drop in the reflectance from unity. In RBCO superconductors, the reflectance is less than one well below the gap due to insufficient screening by the electrons and absorption by phonons (or possibly some other process). The reflectance minima, due to the coupled plasmon-phonon modes discussed earlier, can be comparably deep in both the normal and superconducting states. This causes a drop in the ratio R_S/R_N to near unity. In Fig. 2 we see such minima at 155 cm^{-1} and from 275 to 300 cm^{-1} for the Y and Gd samples. That the ratio recovers to a value well above unity at a higher frequency is evidence that the sudden drop in the ratio at these frequencies is not the free carrier absorption edge. In particular, the band of phonons around 300 cm^{-1} probably prevents the ratio from returning to its maximum value until well above 350 cm^{-1}, so that a value for ω_g of 350 cm^{-1} or greater is plausible. Such values would bring the spectroscopic energy gap into agreement with the tunneling results and characterizes the material as very strong coupled (i.e. $2\Delta/k_B T_c > 4$).

Conclusions

We have studied the infrared reflectance of several high T_c superconductors and find similar phonon and plasma frequencies for each. Slight differences in the phonon frequencies are attributed to structural deformations which result from the different sizes of the rare earth ions.

The very low plasma frequencies give rise to coupled plasmon-phonon modes. These coupled modes may be partially responsible for the remarkably high transition temperatures of these materials.

Finally, we observe enhanced reflectivity in the superconducting state to frequencies up to 350 cm^{-1} and interpret this as evidence for an energy gap frequency near or above 300 cm^{-1}. Strong phonons and coupled plasmon-phonon modes cause the reflectance ratio to drop to near unity at several low frequencies, giving the appearance of a reduced gap.

References

1. D. A. Bonn, J. E. Greedan, C. V. Stager, T. Timusk, M.G. Doss, S. L. Herr, K. Kamaras and D. B. Tanner, Phys. Rev. Lett., 58, 2249 (1987).

2. S. Perkowitz, G. L. Carr, B. Lou, S. S. Yom, R. Sudharsanan and D. S. Ginley, to appear in Solid St. Comm.

3. J. M. Wrobel, S. Wang, S. Gygax, B. P. Clayman and L. K. Peterson, Phys. Rev. B36, 2368 (1987).

4. J. R. Kirtley, R. T. Collins, Z. Schlesinger, W. J. Gallagher, R. L. Sandstrom, T. R. Dinger and D. A. Chance, Phys. Rev. B35, 8846 (1987).

5. J. Moreland, A. F. Clark, L. F. Goodrich, H. C. Ku and R. N. Shelton, Phys. Rev. B35, 8711 (1987).

6. M. F. Crommie, L. C. Bourne, A. Zettl, M. L. Cohen and A. Stacy, Phys. Rev. B35, 8853 (1987).

7. C. Kittel, "Introduction to Solid State Physics", 5th ed. (Wiley, New York, 1976) p. 31.

8. B. B. Varga, Phys. Rev. 137, A1896 (1965).

9. S. Perkowitz, to be submitted.

10. M. L. Cohen, in "Superconductivity", R. D. Parks, ed., vol. 1 (Marcel-Dekker, New York, 1969) pp. 615-664.

22

Simulation of Crystal Structures by Empirical Atom-Atom Potentials. IV. Interpretation of the Raman and Infrared Spectra of $La_{2-x}M_xCuO_4$

MICHEL EVAIN and MYUNG-HWAN WHANGBO   Department of Chemistry, North Carolina State University, Raleigh, North Carolina 27695

JOHN R. FERRARO and JACK M. WILLIAMS   Chemistry and Materials Science Divisions, Argonne National Laboratory, Argonne, Illinois 60439

1. EMPIRICAL ATOM-ATOM POTENTIALS

La_2CuO_4 contains layers of composition CuO_4, which are made up of corner sharing CuO_6 octahedra, and has La^{3+} cations at the 9-coordination sites in between adjacent CuO_4 layers.[1] When such La^{3+} sites are randomly replaced by alkaline earth cations M^{2+} (M = Ba, Sr), one obtains the high temperature superconductors $La_{2-x}M_xCuO_4$ ($x \simeq 0.1 - 0.2$; $T_c \simeq 30-40K$).[2-15] The crystal structure of La_2CuO_4 is orthorhombic below 533K, but tetragonal above 533K.[16] This tetragonal to orthorhombic (T → O) transition is not a Peierls distortion.[17] $La_{2-x}M_xCuO_4$ is tetragonal at room temperature, but undergoes a T → O distortion at a lower temperature (i.e., at 200-215K for $La_{2-x}Sr_xCuO_4$[18-20] and at 180K for $La_{2-x}Ba_xCuO_4$[21]).

Empirical atom-atom potentials provide a practical tool by which to estimate free energies of crystalline solids.[22-24]

Recently, atom-atom potential calculations on La_2CuO_4 showed[25] that the T $\rightarrow$ O distortion originates from the ionic interactions associated with the La^{3+} cations. For ions i and j separated by the distance r_{ij} with the charges q_i and q_j, respectively, the interaction potential W_{ij} can be written as

$$W_{ij} = q_i q_j / r_{ij} + B_{ij} \exp(-r_{ij}/\rho_{ij}) - C_{ij}/r_{ij}^6 \qquad (1)$$

where the first, second and third terms refer to the Coulomb, nonbonded repulsion and van der Waals energies, respectively. The constants B, ρ and C are adjustable parameters to be derived on the basis of experimental data. Under the approximation that the B, ρ and C values between different kinds of ions are related to those between identical ions as

$$B_{ij} = \sqrt{B_{ii} B_{jj}}$$
$$1/\rho_{ij} = (1/\rho_{ii} + 1/\rho_{jj})/2 \qquad (2)$$
$$C_{ij} = \sqrt{C_{ii} C_{jj}}$$

it is sufficient to derive the B, ρ and C values for $O^{2-}...O^{2-}$, $Cu^{2+}...Cu^{2+}$ and $La^{3+}...La^{3+}$ pairs to be able to define potentials for all pairs of ions in La_2CuO_4.[25] The B, ρ and C values for these three pairs of ions, derived by using the program WMIN[26] so as to reproduce the crystal structures of CuO[27] and La_2O_3,[28,29] describe the crystal structures of orthorhombic and tetragonal La_2CuO_4 very well.[25] Atom-atom potential calculations based upon these B, ρ and C values also show[25] that orthorhombic La_2CuO_4 is slightly more stable than tetragonal La_2CuO_4, in good agreement with experiment.

In the present work, we carry out atom-atom potential calculations by using the program WMIN[26] to obtain the zone center (k=0) vibrational frequencies of tetragonal La_2CuO_4 and their corresponding normal modes of vibrations. Our main objective is to help interpret the Raman and infrared (ir) spectra of $La_{2-x}M_xCuO_4$, which is tetragonal at room temperature. In our WMIN calculations, the frequencies computed with and without relaxing the unit cell and atom positional parameters are found to be nearly the same. Symmetry assignments of the normal modes of vibrations are easier with those calculations without relaxing the opitimum geometry, and hence we present those results in the following.

2. RAMAN ACTIVE MODES

The first column of Table 1 lists the symmetries and frequencies of the four Raman active vibrations (two A_{1g} and two E_g modes) calculated for tetragonal La_2CuO_4. In other columns of Table 1, we summarize the Raman frequencies observed[30-34] for $La_{2-x}M_xCuO_4$. The normal modes of the four vibrations are depicted in Figure 1, which shows that all the Raman modes involve only the movements of the axial oxygen atoms of the CuO_6 octahedra and the La atoms. These atoms move along the c-axis direction in the A_{1g} modes, but along the direction perpendicular to the c-axis in the E_g modes. In each E_g mode, two orthogonal, but equivalent, movements of the axial oxygen and the La atoms can be taken along the a- and b-axis directions. For simplicity, only one of the two is shown for each E_g mode in Figure 1. The

Table 1. Calculated Raman Frequencies of Tetragonal La$_2$CuO$_4$ and Observed Raman Frequencies of La$_{2-x}$M$_x$CuO$_4$[a,b]

La$_2$CuO$_4$		La$_{2-x}$M$_x$CuO$_4$			
Present work[c]	Ref. 30	Ref. 31[d]	Ref. 32	Ref. 33	Ref. 34
73 (1E$_g$)	95 (E$_g$)	149 (E$_g$) [149]	152	138 (A$_{1g}$)	100 (A$_{1g}$)
	144 (A$_{1g}$)				180 (E$_g$)
225 (1A$_{1g}$)	220 (E$_g$)	226 (A$_{1g}$) [229]		211 (E$_g$)	
				270 (?)	
375 (2E$_g$)	375	367 (E$_g$) [370]	366		380 (A$_{1g}$)
643 (2A$_{1g}$)	421 (A$_{1g}$)	431 (A$_{1g}$) [425]	430	412 (A$_{1g}$)	

a. Frequencies are in units of cm^{-1}.   b. The symmetry of each Raman mode, when available, is given in parenthesis.   c. Normal modes with an identical symmetry but with different frequencies are distinguished by the numbers in front of the symmetry labels.   d. The frequencies in square brackets are simulated values by assuming a set of stretching force constants.

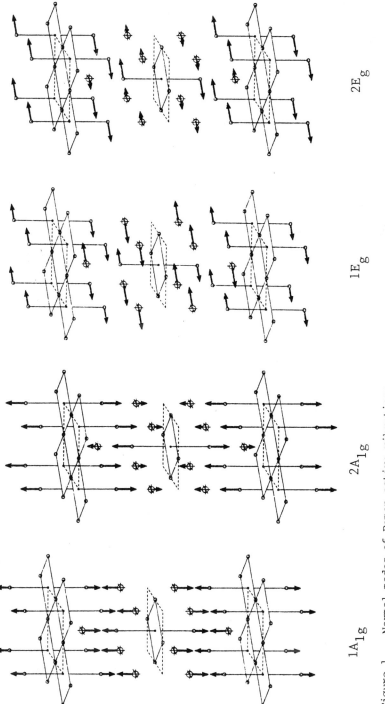

$1A_{1g}$ $2A_{1g}$ $1E_g$ $2E_g$

Figure 1.   Normal modes of Raman active vibrations

sequence of our calculated Raman modes is in agreement with that simulated by Brun et al.[31] The calculated Raman frequencies are also in reasonable agreement with those observed by Brun et al.[31]

3. IR ACTIVE MODES

The first column of Table 2 summarizes the symmetries and frequencies of the seven ir active vibrations (three A_{2u} and four E_u modes) calculated for tetragonal La_2CuO_4. The ir frequencies observed[30,31,33,35-39] for $La_{2-x}M_xCuO_4$ are also listed in other columns of Table 2. The normal modes of vibrations for the ir frequencies are depicted in Figure 2. All the ir modes involve the movements of the equatorial oxygen atoms of the CuO_6 octahedra and the Cu atoms. All the atoms move along the c-axis direction in the A_{2u} modes, but along the direction perpendicular to the c-axis in the E_u modes. Experimentally, the $3E_u$ mode is the only frequency consistantly observed by all studies. The six ir frequencies are observed in some studies, but not observed in other studies, thereby making it difficult to assign the ir frequencies. Nevertheless, our calculated ir frequencies and their sequence are in reasonable agreement with those simulated by Brun et al.[31]

In our atom-atom potential calculations, the $3E_u$ mode is calculated to have an imaginary frequency. In general, a vibration with an imaginary frequency means that the lattice is soft toward such a mode of vibration. The imaginary frequency for the $3E_u$ mode calculated in our study should be a direct consequence

Table 2. Calculated IR Frequencies of Tetragonal La_2CuO_4 and Observed IR Frequencies of $La_{2-x}M_xCuO_4$[a,b]

La₂CuO₄			La₂₋ₓMₓCuO₄				
Present work[c]	Ref. 30	Ref. 31[d]	Ref. 33	Ref. 35	Ref. 36	Ref. 37	Refs. 38-39
98 ($1E_u$)		[132] (E_u)					
143 ($1A_{2u}$)	150	150 (A_{2u}) [140]					
	247 (?)					255	
220 ($2A_{2u}$)	336	337 (E_u) [340]		342	337		
	356 (?)			356			
247 ($2E_u$)	445 (?)	[406] (A_{2u})	445				
i ($3E_u$)[e]	500	500 (E_u) [507]	490	496	500	505	505-520
	560 (?)		560				

Table 2. (Continued)

La$_2$CuO$_4$		La$_{2-x}$M^M_xCuO$_4$						
Present work[c]	Ref. 30	Ref. 31[d]	Ref. 33	Ref. 35	Ref. 36	Ref. 37	Refs. 38–39
824 (4E$_u$)		[620] (A$_{2u}$)					
853 (3A$_{2u}$)		[646] (E$_u$)					

[a.] Frequencies are in units of cm^{-1}.   [b.] The symmetry of each Raman mode, when available, is given in parenthesis.   [c.] Normal modes with an identical symmetry but with different frequencies are distinguished by the numbers in front of the symmetry labels.   [d.] The frequencies in square brackets are simulated values by assuming a set of stretching force constants.   [e.] Imaginary frequency.

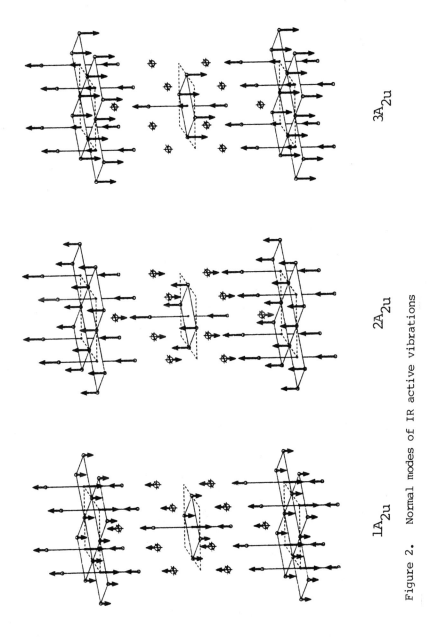

3A$_{2u}$

2A$_{2u}$

1A$_{2u}$

Figure 2.   Normal modes of IR active vibrations

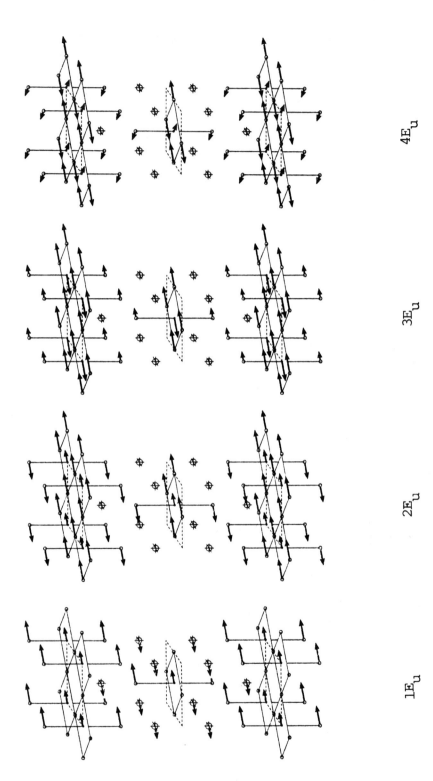

$1E_u$ $2E_u$ $3E_u$ $4E_u$

Figure 2.   Normal modes of IR active vibrations (continued)

of our empirical atom-atom potentials, which do not adequately represent the potential well associated with a bond angle change around an atom. In the $3E_u$ mode of vibration, all the Cu atoms move in one direction, while all the oxygen atoms move in the opposite direction, in such a way that these atom movements lead to the bond angle changes around the Cu and oxygen atoms.

4. RAMAN AND IR INACTIVE MODE

One mode of vibration with the symmetry B_{2u}, shown in Figure 3, is not Raman or ir active. This vibration involves only the out-of-phase movements of the equatorial oxygen atoms of the CuO_6 octahedra, and its vibrational frequency is calculated to be 158 cm^{-1}.

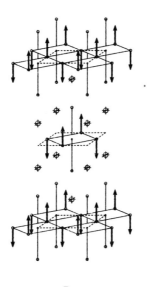

B_{2u}

Figure 3.   Normal mode of a Raman and IR inactive vibration

5. CONCLUDING REMARKS

Our atom-atom potential calculations lead to the four Raman frequencies that are in reasonable agreement with those observed for $La_{2-x}M_xCuO_4$. As for the ir frequencies of $La_{2-x}M_xCuO_4$, only one frequency around 500 cm^{-1} is consistently found in a number of studies. Further experimental studies are necessary to properly assign the ir spectra of $La_{2-x}M_xCuO_4$. On the theoretical side, there is a need to improve the atom-atom potentials by including potentials that properly describe bond angle changes.

ACKNOWLEDGMENT

Work at North Carolina State University and Argonne National Laboratory were supported by the U.S. Department of Energy, Office of Basic Energy Sciences, Division of Materials Sciences under Grant DE-FG05-86-ER45259 and under Contract W31-109-ENG-38, respectively. We express our appreciation for computing time made available by DOE on the ER-Cray X-MP computer. We are grateful to Dr. W. R. Busing for making his WMIN program available to us.

REFERENCES

1. Grande, B.; Müller-Buschbaum, Hk.; Schweizer, M. Z. Anorg. Allg. Chem., **1977**, <u>424</u>, 120.

2. Bednorz, J.G.; Müller, K.A. Z. Phys. B. COndensed Matter, **1986**, <u>64</u>, 189.

3. Bednorz, J.G.; Takashige, M.; Müller, K.A. Europhys. Lett., **1987**, <u>3</u>, 379.

4. Takagi, H.; Uchida, S.; Kitazawa, K.; Tanaka, S. Jpn. J.
    Appl. Phys., Part 2, **1987**, 26, L123

5. Uchida, S.; Takagi, H.; Kitazawa, K.; Tanaka, S. Jpn. J.
    Appl. Phys., Part 2, **1987**, 26, L1.

6. Cava, R.J.; van Dover, R.B.; Bartlogg, B.; Rietmann, E.A.
    Phys. Rev. Lett., **1987**, 58, 408.

7. Chu, C.W.; Hor, P.H.; Meng, R.L.; Gao, L.; Huang, Z.J.;
    Wang, Y.Q., Phys. Rev. Lett., **1987**, 58, 405.

8. Chu, C.W.; Hor, P.H.; Meng, R.L.; Gao, L.; Huang, Z.J.
    Science (Washington, D.C.), **1987**, 235, 567.

9. Jorgensen, J.D.; Schuttler, H.-B.; Hinks, D.G.; Capone, D.W.;
    Zhang, K.; Brodsky, M.B.; Scalapino, D.J. Phys. Rev. Lett.,
    1987, 58, 1024.

10. Tarascon, J.M.; Greene, L.H.; McKinnon, W.R.; Hull, G.W.;
    Gabelle, T.H. Science (Washington, D.C.), **1987**, 237, 1373.

11. Bonne, D.A.; Greedan, J.E.; Stager, C.V.; Timusk, T. Solid
    State Commun., **1987**, 62, 383.

12. Beille, J.; Cabanel, R.; Chaillout, C.; Chevalier, B.;
    Demazeau, G.; Deslandes, F.; Etourneau, J.; Lejay, P.;
    Michel, C.; Provost, J.; Raveau, B.; Sulpice, A.; Tholence, J.-L.;
    Thornier, R. C.R. Seances Acad. Sci. Ser. 2, **1987**,
    304, 1097.

13. Grant, P.M.; Parkin, S.S.P.; Lee, V.Y.; Engler, E.M.; Ramirez, M.L.;
    Lim, G.; Jacowitz, R.D.; Greene, R.L. Phys. Rev. Lett.,
    1987, 58, 2482.

14. Kanbe, S.; Kishio, K.; Kitazawa, K.; Fueki, K.; Takagi, H.;

Tanaka, S. Chem. Lett., **1987**, 547.

15. Kishio, K.; Kitazawa, K.; Sugii, N.; Kanbe, S.; Fueki, K.; Takagi, H.; Tanaka, S., Chem. Lett., **1987**, 635.

16. Longo, J.M.; Raccah, P.M. J. Solid State Chem., **1973**, 6, 526.

17. Whangbo, M.-H.; Evain, M.; Beno, M.A.; Williams, J.M. Inorg. Chem. **1987**, 29, 1829.

18. Cava, R.J.; Santoro, A.; Johnson, D.W.; Rhodes, W.W. Phys. Rev. B: Condens. Matter, **1987**, 35, 6716.

19. Moret, R.; Pouget, J.P. Collin, G. Europhys. Lett., **1987**, in press.

20. Geiser, U.; Beno, M.A.; Schultz, A.J.; Wang, H.H.; Allen, T.J.; Monaghan, M.R.; Williams, J.M. Phys. Rev. B: condens. Matter, **1987**, 35, 6721.

21. Paul, D. Mak; Balakrishnan, G.; Bernhoeft, N.R.; David, W.I.F.; Harrison, W.T.A. Phys. Rev. Lett., **1987**, 58, 1976.

22. Stoneham, A.M.; Harding, J.H. Ann. Rev. Phys. Chem., **1986**, 37, 53.

23. "Computer Simulation of Solids"; Catlow, C.R.A.; Mackrodt, W.C., eds.; Springer-Verlag: New York, 1982.

24. Williams, D.E. Topics in Current Physics, **1981**, 26, 3.

25. Evain, M.; Whangbo, M.-H.; Beno, M.A.; Geiser, U.; Williams, J.M. J. Am. Chem. Soc. (in press).

26. Busing, W.R. "WMIN, A computer program to model molecules and crystals in terms of potential energy functions," Oak Ridge National Laboratory, ORNL-5497: Oak Ridge, 1981.

27. Asbrink, S.; Norrby, L.-J. Acta Cryst. **1970**, B26, 8.

28. Pauling, L. Z. Krist, **1929**, 69, 415.

29. Wyckoff, R.W.G. "Crystal Structures, Vol. 2", Wiley: New York, 1964; 2nd ed.; Chap. V, Section A.

30. Maroni, V.A.; Brun, T.O.; Johnson, S.A.; Grimsditch, M.; Loong, C.K.; Melendres, C.A. Presented at the A.C.S. Symposium on Inorganic Superconductors, New Orleans (1987).

31. Brun, T.; Grimsditch, M.; Gray, K.E.; Bhadra, R.; Maroni, V. Submitted to Phys. Rev. Lett. (1987).

32. Sugai, S.; Sato. M.; Hosoya, S.; Jap. J. Appl. Phys. **1987**, 26, L495.

33. Copic, M.; Milhailovic, D.; Zgonik, M.; Prester, M.; Biljakovic, K.; Orel. B.; Brnicevic, N. Submitted to Sol. State Comm. (1987).

34. Blumenroeder, S.; Zirngiebl, E.; Thompson, J.D.; Killough, P.; Smith, J.L.; Fisk, Z. Submitted (1987).

35. Bonn, D.A.; Greedan, J.E.; Stager, C.V.; Timusk, T.; Doss, M.; Herr, S.; Kamaras, K.; Porter, C.; Tanner, D.B.; Tarascon, J.M.; McKinnon, W.R.; Greene, L.H. Submitted (1987).

36. Oh-ishi, K.; Kikuchi, M.; Syono, Y.; Hiraga, K.; Morioka, Y. Submitted (1987).

37. Sawada, H.; Saito, Y.; Iwazumi, T.; Yoshizaki, R.; Abe Y.; Matsura, E. Jap. J. Appl. Phys., **1987**, 26, L426.

38. Ohbayashi, K.; Ogita, N.; Udagawa, M.; Aoki, Y.; Maeno, Y.; Fujita, T. Jap. J. Appl. Phys., **1987**, 26, L420.

39. Stavola, M.; Cava, R.J.; Rietman, E.A. Phys. Rev. Lett., 58, 1571.

23

Investigation of Y-Ba-Cu-O Superconducting Materials by Positron Annihilation Lifetime Spectroscopy

A. J. HILL, F. H. COCKS, U. M. GOESELE, P. L. JONES, and **T. Y. TAN** Department of Mechanical Engineering and Materials Science, Duke University, Durham, North Carolina 27706

A. I. KINGON Department of Materials Science and Engineering, North Carolina State University, Raleigh, North Carolina 27695-7907

INTRODUCTION

Positron trapping at vacancies and vacancy complexes is a well established phenomenon in most metals and in some semiconductors.[1-4] In addition, Groznov et. al.[5] measured positron trapping at defects in the superconducting oxides, $BaPb_{1-x}Bi_xO_3$, and most recently, Jean et. al.[6] reported results indicating positron trapping at oxygen vacancies in $YBa_2Cu_3O_{6+\delta}$ samples. The lifetime of a positron trapped at a defect site is greater than the lifetime of a quasi-free positron in the bulk lattice due to the decrease in electron density at the defect site. The amount of increase in the defect lifetime over the bulk lifetime can yield some idea of the relative size of the defect.

The crystal structure of the superconducting $YBa_2Cu_3O_{7-x}$ material has been identified as an orthorhombic perovskite-derivative structure with oxygen vacancies present along non-intersecting Cu-O chains in the lattice.[7,8] These Cu-O chains are located in the Cu plane between the Ba ions and are present due to ordered oxygen vacancies in the material. Removal or addition of oxygen to this Cu plane causes changes in the superconducting properties. In fact, the existence and specific value of T_c in the Y-Ba-Cu-O system depends critically on the concentration and distribution of oxygen vacancies in the lattice as evidenced by the direct dependence of annealing atmosphere and cooling procedures on the occurrence of a high T_c superconducting phase.[7-12] Thus, because positrons are a probe for crystal defects such as vacancies, the lifetime technique should be a useful tool in the study of the effect of oxygen vacancies on the superconducting transition of $YBa_2Cu_3O_{7-x}$.

EXPERIMENTAL

Positron measurements were made on two sets of $YBa_2Cu_3O_{7-x}$ samples prepared at North Carolina State University and at Duke University. The NCSU samples were prepared using $BaCO_3 + CuO + Y_2O_3$ starting materials which were mixed and reacted at 930° C for 12 hours in an oxygen atmosphere. The material was then quenched and pressed into discs and sintered at 950° C for 10 hours in O_2 and slow cooled in O_2. The Duke samples were prepared using $BaCO_3 + CuO + Y_2O_3$ starting materials which were mixed and reacted at 950° C for 16 hours in O_2. The material was cooled to room temperature,crushed, pressed into discs and sintered at 875°C for 4 hours, cooled to 500° C at 3°C/min. in O_2 and subsequently air-quenched to room temperature. Both sets of samples were measured for the dependence of resistance on temperature by the four point probe method and were found to be superconducting at 77K. The NCSU samples were measured continuously as a function of temperature and found to have $T_c=91K$.

The positron annihilation spectra were obtained using a fast-fast coincidence system with a timing resolution of 225 psec based on the FWHM of ^{60}Co with the energy windows set for ^{22}Na events. Measurements were made in the normal sandwich configuration with the positron source consisting of 30μCi of ^{22}Na deposited on and sealed between .2mil Ti foil. The 77K measurements were performed by sealing the sample in a desiccated chamber prior to immersing in a liquid nitrogen bath. Each spectrum contained at least 10^5 counts and were modeled using the computer program PFPOSFIT.[13]

RESULTS AND DISCUSSION

The results of the NCSU and Duke samples are shown in Table 1. The Duke samples at 298K yielded spectra that were consistently modeled by a three component fit with all lifetimes unconstrained. The remaining spectra could not be consistently modeled by an unconstrained three component fit, thus one lifetime was constrained and the analyses were repeated until consistent results were obtained. For these spectra, the intermediate component lifetime was fixed and varied over a range from .300 nsec to .600 nsec by 10 psec increments in order to get the best fit and best standard deviations of the resolved lifetimes and intensities. This partially constrained three component fit provides consistent representation of the annihilation spectra. The mean variance of the fits was 1.09.

The shortest lifetime component can be attributed to the annihilation of quasi-free positrons in the bulk of the material based on its lifetime and intensity values. The range of τ_1, .175 nsec to .187 nsec, is characteristic of bulk annihilation lifetimes measured in metals and semiconductors,[14] and the intensity value indicates that the majority of annihilations are taking place in the bulk as expected. The longest lifetimes and intensities (τ_3, I_3) are attributable to annihilations in the source material and at the sample surface and probably to ortho-positronium annihilations.

Table 1. Lifetime and intensity values for Duke and NCSU samples.

	$T°$	Lifetimes (nsec)			Intensities (%)		
		τ_1	τ_2	τ_3	I_1	I_2	I_3
DUKE	298K	.187±.005	.459±.029	2.370 ±.09	79.2±2.4	15.3±2.4	5.5±.2
	77K	.175±.010	.430fixed	1.480±.11	71.1±1.9	23.2±2.5	5.7±.6
NCSU	298K	.180±.001	.480fixed	2.663±.64	96.7±.4	2.5±.5	.74±.1
	77K	.183±.003	.430fixed	1.947±.45	96.8±.3	2.8±.3	.47±.1

τ_2 is considerably larger than τ_1 indicating an annihilation environment with a much lower electron density than the bulk state. The predominant defects in the Y-Ba-Cu-O material are the oxygen vacancies. The removal of an oxygen atom in the crystal structure creates an anion vacancy, thus the vacancy would have a net positive charge which would not tend to attract the positron (although it should be noted that association of the anion vacancy with a particular cation could effectively neutralize the vacancy locally, making it attractive to the positron). If however, the positron were trapped in the anion vacancy, the low electron density in the trap would increase the positron lifetime substantially over that of the bulk or that expected in an neutral monovacancy. Table 2 contains comparative values of bulk and

defect lifetimes in a variety of materials for annihilations in various types of
defects. It can be seen from Table 2 that for the present work, a τ_2/τ_1 value
of 2.486 (.450/.181) is consistent with lifetime values of positrons trapped at
cation vacancies or vacancy agglomerates but is considerably larger than
would be expected if the defect were an uncharged monovacancy. These
results indicate that the defect site possibly could be either a charged vacancy
or a vacancy complex formed by the ordered vacancies in the Cu-O
connecting chains in the $YBa_2Cu_3O_{7-x}$ structure.

Table 2. Bulk lifetime and defect lifetime values in various materials.

Material	Type of defect	bulk lifetime τ_1 (nsec)	defect lifetime τ_2 (nsec)	τ_2 / τ_1	REF.
Fe	monovacancy	.160	.175	1.094	15
GaAs	monovacancy	.233	.255	1.094	3
Si	monovacancy	.220	.226	1.209	2
pGe	monovacancy	.230	.290	1.261	4
Fe	vacancy cluster	.160	.300	1.875	15
Si	quadrivacancy	.220	.430	1.954	2
NaCl	cation vacancy	.217	.454	2.092	16
$BaPb_9Bi_1O_3$	cation vacancy	.178	.390	2.191	5

The intensity of the defect component, I_2, varies considerably between the
Duke samples and the NCSU samples.   Beyers et. al.[12] have concluded from
TEM and neutron diffraction studies that during the preparation of the
superconducting Y-Ba-Cu-O material, quench rate determines the oxygen
incorporation and the oxygen vacancy ordering in the Cu-O connecting
chains. Because the materials were not prepared under identical conditions,
the differences in the I_2 values are not surprising. The range of I_2 (2.5% to
23.2%), is similar to the range of I_2 measured by Groznov et. al.[5] in the
superconducting $BaPb_{1-x}Bi_xO_3$ material as a function of Bi concentration
(x). In the Groznov study, as x was varied from 0 to .4, the defect lifetime
(τ_2) and intensity (I_2) also varied as a function of x. The lifetime ranged
from .36 nsec to .58 nsec and the intensity varied from 27% to 4%. Thus,
because stoichiometry had a marked effect on the defect component lifetime

and intensity in the perovskite superconducting Ba-Pb-Bi-O material, it is probable that the variations in I_2 seen between samples in this study are due to stoichiometry.

The Jean et. al.[6] study of positron annihilation in $YBa_2Cu_3O_{6+\delta}$ presents spectra measured over a temperature range from 10K to 293K. The spectra were modeled by a three component fit yielding lifetimes $\tau_1 \sim 139$ psec, $\tau_2 \sim 210$ psec and $\tau_3 \sim 2.5$ nsec. The defect component lifetime, τ_2, decreases abruptly by approximately 14 psec (with decreasing temperature), at $T_c = 90K$ in superconducting samples ($\delta = .8$) and remains linear over the temperature range for nonsuperconducting samples ($\delta = 0$). The defect component intensity, I_2, increases abruptly by approximately 6%-7% (with decreasing temperature), at $T_c = 90K$ and remains approximately constant over the temperature range for nonsuperconducting samples. I_2 values in the superconducting samples ranged from ~25% to ~35% in the Jean et. al.[6] study.

A comparison of the Jean et. al.[6] study with the present study yields several significant points. The average defect lifetime value for the present investigation is .450 nsec as compared with the value $\tau_2 \sim .210$ nsec for the Jean et. al.[6] study. The spectra of the present study were reanalysed in an attempt to discern τ_1 and τ_2 values similar to the Jean et. al.[6] results; however, the spectra could not be consistently fit in this manner. The I_2 values also differ between the two studies, but it is thought that these differences may be attributable to sample preparation techniques. Based on the initial data presented in the present study, a decrease in τ_2 of approximately 30 psec to 50 psec occurs between the normal and superconducting state. In addition, the intensity of the defect component in the Duke samples increased by approximately 8% between the normal and superconducting state. The results of this study support the general trends of I_2 and τ_2 behavior as a function of temperature going from the normal to the superconducting state presented in the research of Jean et. al.[6] The τ_2/τ_1 ratio of the Jean et. al.[6] study is (210/139) 1.51, which is somewhat lower than the ratio calculated for the present study. The τ_2/τ_1 ratio of the present study indicates that the predominant defect trapping site in this material is possibly a charged vacancy or a vacancy complex.

CONCLUSIONS

The results of the present investigation and the Jean et. al.[6] study demonstrate that the positron annihilation lifetime technique is a potentially useful tool for elucidating the superconducting mechanism in the 1-2-3 compound. Because no difference in PAL spectra was observed in the more traditional metal alloy superconductors,[17] the existence of a difference in the PAL spectra upon the superconducting transition of the 1-2-3 compound could mean either an electronic transition or a phase transition (in the sense of the oxygen stoichiometry and ordering), may have taken place in association with the onset of superconductivity in the material. In view of the PAL spectral differences in the Duke/NCSU/Jean sample results, one needs to first address extraneous factors, such as sample preparation,which may be the cause of these differences. Additional positron studies on carefully prepared samples as a function of temperature and oxygen stoichiometry are planned in order to help clarify the role of oxygen vacancies in the occurrence of the superconducting transition in the $YBa_2Cu_3O_{7-x}$ material.

ACKNOWLEDGMENTS

One of the authors (A.J.H.) would like to thank the National Science Foundation and the Army Research Office for financial support.

REFERENCES

[1] A. Seeger, J. Phys. F: Metal Phys. **3**, 248 (1973).
[2] S. Dannefaer, B. Hogg, and D. Kerr, in "Proc. 13th Internat. Conf. on Defects in Semiconductors," eds. L. C. Kimerling and J. M. Parsey, Jr. (The Metall. Soc. of AIME, Warrendale, PA, 1985), p. 255.
[3] M. Stucky, C. Corbel, B. Geffroy, P. Moser and P. Hautojarvi, in "Defects in Semiconductors," ed. H. J. von Bardeleben (Trans Tech Publ. Ltd., Switzerland, 1986) p. 265.
[4] C. Corbel, P. Moser, and M. Stucky, Ann. Chim. Fr. **8**, 733 (1985).
[5] I. N. Groznov, S. P. Ionov, I. B. Kevdina, V. S. Lubimov, E. F. Makarov, O. V. Marchenko, N. N. Mikhailov, V. A. Onishchuk, and V. P. Shantarovich, Phys. Stat. Sol. (b) **123**,183 (1984).
[6] Y. C. Jean, S. J. Wang, H. Nakanishi, W. N. Hardy, M. E. Hayden, R. F. Kiefl, R. L. Meng, H. P. Hor, J. Z. Huang, and C. W. Chu, Phys. Rev. B **36**, 3994 (1987).
[7] R. L. Cava, B. Batlogg, R. B. van Dover, D. W. Murphy, S. Sunshine, T.

Siegrist, J.P. Remeika, E. A. Rietman, S. Zahurak, and G. P. Espinosa, Phys. Rev. Lett. **58**, 1676 (1987).

[8]F. Beech, S. Miraglia, A. Santoro, and R. S. Roth, Phys. Rev. B **35**, 8778 (1987).

[9]S. B. Qadri, L. E. Toth, M. Osofsky, S. Lawrence, D. U. Gubser, and S. A. Wolf, Phys. Rev. B **35**, 7235 (1987).

[10]P. H. Hor, R. L. Meng, Y. Q. Wang, L. Gao, Z. J. Huang, J. Bechtold, K. Forster, and C. W. Chu, Phys. Rev. Lett. **58**, 1891 (1987).

[11]J. S. Tsai, Y. Kubo, and J. Tabuchi, Phys. Rev. Lett. **58**, 1979 (1987).

[12]R. Beyers, G. Lim, E. M. Engler, V. Y. Lee, M. L. Ramirez, R. J. Savoy, R. D. Jacowitz, T. M. Shaw, S. La Placa, R. Boehme, C. C. Tsuei, Sung I. Park, M. W.Shafer, and W. J. Gallagher, Appl. Phys. Lett. **51**, 614

[13]W. Puff, Comput. Phys. Commun. **30**, 359 (1983).

[14]H. Weisberg and S. Berko, Phys. Rev. **154**, 249 (1967).

[15]A. Vehanen, P. Hautojarvi, J. Johansson, J. Yli-Kauppila, and P. Moser, Phys. Rev. B **25**, 762 (1982).

[16]W. Brandt and J. Reinheimer, Phys. Lett. A **35**, 109 (1971).

[17]I. Ya. Dekhtyar, Physics Report (Section C of Physics Letters) **9**, 243 (1974).

24

Magnetic Anomalies in the Rare Earth Oxide Superconductors $GdBa_2Cu_3O_{7-x}$ and $YbBa_2Cu_3O_{7-x}$

WILLIAM E. HATFIELD, BRIAN R. ROHRS, MARTIN L. KIRK, JEFFREY H. HELMS, HYEKYEONG RO, and ERIC J. WILLIAMSEN   University of North Carolina at Chapel Hill, Chapel Hill, North Carolina 27599

The discovery of high T_c superconductivity in $YBa_2Cu_3O_{7-x}$[1,2] and this group's interest in the magnetic properties of rare earth compounds led to the study of paramagnetic analogs. Anomalous magnetic behavior in selected samples of two of these systems, $GdBa_2Cu_3O_{7-x}$ and $YbBa_2Cu_3O_{7-x}$ is reported here. A spin - glass like transition is present at temperatures below T_c. The glassy behavior in an annealed pellet of $GdBa_2Cu_3O_{7-x}$ is shown in Figure 1. The flat regions in the diamagnetic shielding curve show flux trapping upon reversing the temperature and cooling in a field, proof that the sample is still a superconductor. The anomalous behavior observed in the diamagnetic shielding shows a time decay over a period of two weeks (Fig. 2) resulting in an increasingly broadened transition. After three weeks, the same sample displays typical superconductor behavior, however, a small residual componant of the previous phase results in a slight decrease in the magnitude of the diamagnetic signal at 50 K. It is believed this transient glassy behavior is due to the sample being subjected to numerous thermal and magnetic cycles. The fact that subsequent samples lost this behavior after one series of such cycles confirms the metastability of this glassy state.

The Meissner signal for the sample showing the anomalous behavior and for the same sample after three weeks are essentially identical and show ~25% bulk diamagnetism. This is typical of a porous, granular superconducting composite.[3,4] The magnetic susceptibility of a powdered sample of $GdBa_2Cu_3O_{7-x}$ (Fig. 3) is qualitatively

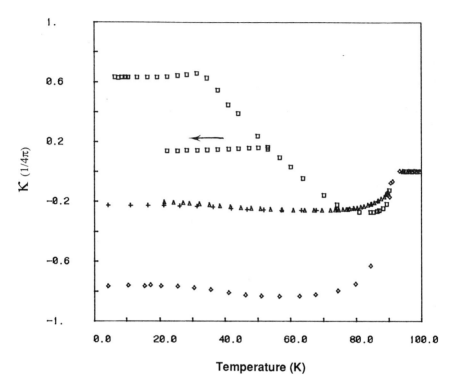

Temperature (K)

Figure 1. Volume susceptibility of $GdBa_2Cu_3O_{7-x}$ showing glassy behavior. Diamagnetic shielding (zero field cooled) signal of original sample ☐ , arrow demonstrates flux trapping upon cooling at 52°K. Meissner (field cooled) signal of original sample Δ . Diamagnetic shielding signal of same sample after three weeks ◊ showing ~85% total flux exclusion. Meissner signal after three weeks + . All susceptibilities are corrected for porosity and demagnetization effects.

similar to Figure 1, except that no increase in the susceptibility below T_c is seen. The fact that the diamagnetic shielding is less than the previous amount of 75% for the pellet may be due to the smaller particle size of the powder.

The $YbBa_2Cu_3O_{7-x}$ sample (Fig. 4) is very similar to $GdBa_2Cu_3O_{7-x}$ since it also shows the glass - like behavior. The Meissner signal is ~10% of the value for total flux exclusion. The magnetization of the $YbBa_2Cu_3O_{7-x}$ specimen (Fig. 5) is typical of a "dirty" superconductor. As the field increases , the magnetization is linear beyond H_{c1} and curves slowly due to the inability of flux to freely penetrate. Upon reversing the field, some of the flux is trapped at defect sites giving rise to the large hysteresis shown.

Several techniques have been used to determine sample purity. X-ray powder diffraction data give no evidence for a separate impurity phase. ICP-AES analysis yields an elemental stoichiometry of Gd$_1$Ba$_{1.922}$Cu$_{2.752}$O$_y$. Both of these techniques are accurate to within 5%. EPR spectroscopy (Fig. 6) at very high gain also yields none of the paramagnetic copper impurities witnessed by others.[5] However, a broad Gd signal is observed at high gain indicating the possibility of a minor impurity phase (less than 0.5%). Perhaps the most convincing evidence lies in the susceptibility measurements. The Meissner signals show none or very small amounts of the large glass-like signal. Furthermore, paramagnetic impurities that give signals of this magnitude do not suddenly

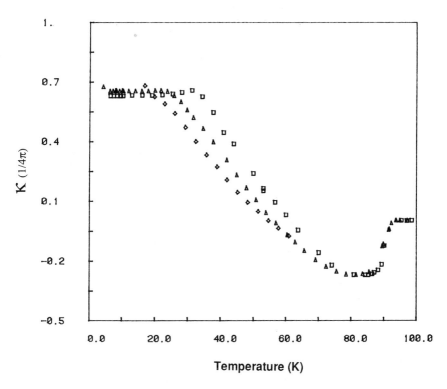

Figure 2. Diamagnetic shielding signal of GdBa$_2$Cu$_3$O$_{7-x}$ demonstrating signal decay over time: original sample □ , one day later Δ , after two weeks ◊ .

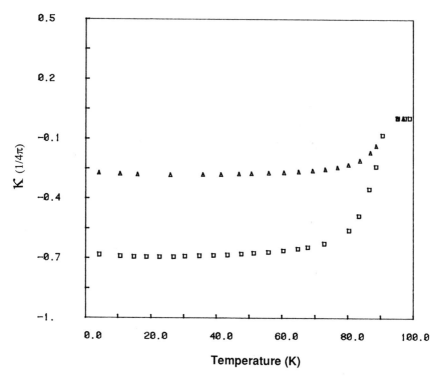

Figure 3. Volume susceptibility of a powdered sample of GdBa$_2$Cu$_3$O$_{7-x}$ showing typical superconductor behavior. The Meissner effect Δ is ~25% and the diamagnetic shielding $\square$ is ~70% of perfect diamagnetism.

dissipate. It can be conclusively stated that the signal is caused by some other phenomenon.

One explanation for this behavior is the occurance of a spin-glass state due to a granular superconductor.[6,7] The basic model involves superconducting grains, each small compared to the penetration depth, which are weakly coupled into closed loops. At finite fields, these clusters are frustrated, i.e., they cannot find a state which minimizes all bond energies simultaneously. Consequently, there are numerous competing ground states with nearly equal energy. Cooling in zero-field, followed by the application of a field at a fixed temperature, produces a metastable state resulting in frustration and spin-

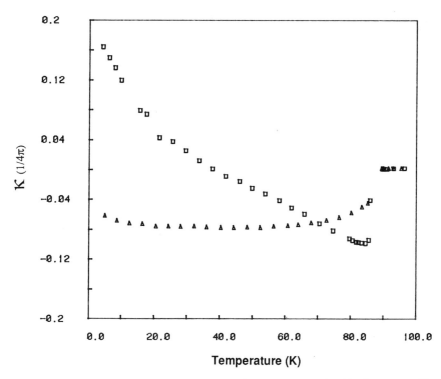

Figure 4. Volume susceptibility of YbBa$_2$Cu$_3$O$_{7-x}$. Diamagnetic shielding signal □ ,
Meissner signal Δ   .

glass like behavior. If the sample is cooled slowly in a fixed field, the clusters are in
equilibrium and a typical Meissner signal results.

This explanation, however, does not account for the large positive increase in the
susceptibility. It accounts only for a decrease in the diamagnetism. Since X-ray powder
diffraction and ICP-AES provide evidence that the impurities are less than 5%, small Gd
impurities may form mictomagnetic clusters resulting in spin-glass formation. [8,9] The ZFC
curve is a result of these clusters being frozen at some temperature T<T$_c$ in a frustrated
manner. Upon warming, the rigid glass softens and the magnetic clusters relax. Thermal
disorder then causes a time averaged magnetization which manifests itself in a decrease

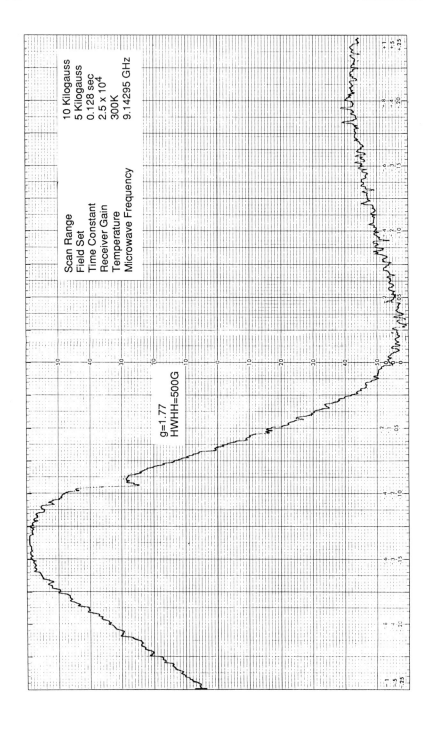

Scan Range   10 Kilogauss
Field Set   5 Kilogauss
Time Constant   0.128 sec
Receiver Gain 2.5×10^4
Temperature   300K
Microwave Frequency   9.14295 GHz

g=1.77
HWHH=500G

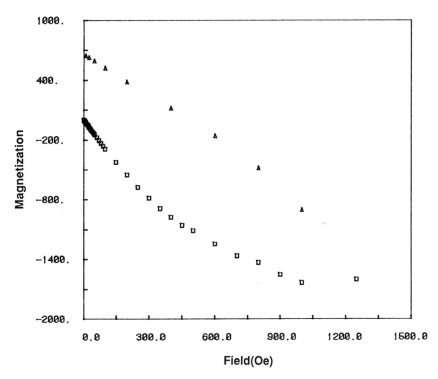

Figure 5. Magnetization of YbBa$_2$Cu$_3$O$_{7-x}$ showing hysteresis due to flux trapping. Increasing field □ , decreasing field △ .

in the susceptibility. In the FC curve, the temperature is gradually decreased and the cluster spins have time to minimize their energy, resulting in no net contribution to the magnetization by the impurity ions. Thermal and magnetic cycling may tend to break up these domain-like clusters. This may explain the transient nature of the observed glassy behavior.

Figure 6. (Previous page)   Room temperature X-band EPR signal of GdBa$_2$Cu$_3$O$_{7-x}$ showing no copper impurities and a small Gd^{2+} signal. The small absorption at g =2.00 is from a DPPH standard. The Gd^{2+} signal is not due to Gd$_2$O$_3$ which has g =1.86 and HWHH =330G.

ACKNOWLEDGMENT

This work was supported in part by the Office of Naval Research.

REFERENCES

1.   C.W. Chu, preprint

2.   M.K. Wu, J.R. Ashburn, C.J. Torng, P.H. Hor, R.L. Meng, L. Gao, Z.J. Huang,
      Y.Q. Wang, and C.W. Chu, *Phys. Rev. Lett.*, *16*, 908 (1987).

3.   Y. Yeshurun, I. Felner, and H. Sompolinsky, *Phys. Rev. B*, *36*, 840 (1987).

4.   J.M. Tarascon, W.R. McKinnon, L.H. Greene, G.W. Hull, and E.M. Vogel, *Phys.
      Rev. B*, *36*, 226 (1987).

5.   R. Jones, M.F. Ashby, A.M. Campbell, P.P. Edwards, M.R. Harrison, A.D. Hibbs,
      D.A. Jefferson, A.I. Kirkland, T. Thanyasiri, and E. Sinn, "Synthesis, Chemistry,
      Electronic Properties ans Magnetism in the Y-Ba-Cu-O Superconductor
      Systems", In "Chemistry of High-Temperature Superconductors", D.L. Nelson,
      M.S. Whittingham, and T.F. George, Eds., ACS Symposium Series 351, p. A1,
      1987.

6.   C. Ebner, and D. Stroud, *Phys. Rev. B*, *31*, 165 (1985).

7.   S.John, and T.C. Lubensky, *Phys. Rev. Lett.*, *55*, 1014 (1985).

8.   P.J. Ford, *Contemp. Phys.*, *23*, 141 (1982).

9.   D. Davidov, K. Baberschke, J.A. Mydosh, and G.J. Nieuwenhuys, *J. Phys. F*, *7*,
      L49   (1977).

25

The Low-Temperature Specific Heat of $YBa_2Cu_3O_{6+\delta}$ Compounds

D. G. HAASE Department of Physics, North Carolina State University, Raleigh, North Carolina 27695-8202

R. VELASQUEZ and A. I. KINGON Department of Materials Science and Engineering, North Carolina State University, Raleigh, North Carolina 27695-8202

Other researchers have reported that the low temperature specific heats of 30 - 40 Kelvin mixed oxide superconductors have a contribution linearly proportional to the temperature.[1-4] It has been hypothesized that normal electrons not participating in the superconducting state are responsible for this contribution. In the present work we have measured the specific heat of the 90 Kelvin superconducting compound $YBa_2Cu_3O_{6+\delta}$ and also find a low temperature contribution linearly proportional to the temperature. However we also find that there is a similar contribution in the insulating analogue to $YBa_2Cu_3O_{6+\delta}$, therefore implying that only a part of the linear specific heat may arise from electrons.

In the past few months several groups have measured the low temperature specific heat of the compounds of the 40 Kelvin family of mixed oxide superconductors. In each of these cases it was reported that the specific heat followed the form $C = \gamma T + \beta T^3$, where βT^3 is the standard Debye lattice contribution ($\beta \sim \Theta_D^{-1/3}$) and the γT term is presumably due to electrons not participating in the superconducting state. In Reference 4 it is pointed out that if the γ term is caused by normal electrons this term should be correlated with the size of the Meissner effect at the superconducting transition. Since the magnitude of the Meissner effect is dependent on the processing of the sample, one would expect that samples from different laboratories would have a range of values of γ, of roughly the same order of magnitude.

Below we give a summary of these specific heat results.

Reference	Compound	γ(mJ/mol-K^2)	Θ_D(K)
1	$La_{1.85}Ba_{0.15}CuO_{4-y}$	~10	(359)
	$La_{1.8}Sr_{0.2}CuO_{4-y}$	~5	(356)
2	$La_{1.85}Sr_{0.15}CuO_{4-y}$	5 ± 1	450
3	$La_{1.8}Ba_{0.2}CuO_{4-y}$	5.4	394
4	$(La_{0.9}Ba_{0.1})_2CuO_{4-y}$	6	330
	$(La_{0.91}Sr_{0.09})_2CuO_{4-y}$	2.8	360

(For the sake of consistency the values in parentheses have been recalculated from those in the reference to allow for a unit cell of seven atoms each having three degrees of freedom.)

The values of γ for the different samples with two exceptions agree very well with each other, perhaps better than one would expect.

We have measured several samples of YBaCuO compounds produced here and at other laboratories. The γ terms in these compounds were not the same, as found in the 40 Kelvin family, but had values of γ ranging from 5 to 20 mJ/mol-K^2. In an effort to better understand the source of the γ term we decided to measure the specific heat of a superconducting specimen and a sample of the non-oxygenated insulating compound, both produced from the same pill.

Our samples were produced by a mixed powder preparation scheme as previously reported.[5] $BaCO_3$ (Fisher, 99.7%; 0.2% $SrCO_3$), CuO (Alfa Products, ACS Grade), and Y_2O_3 (Aesar, 99.9%) were mixed and reacted in air in Al_2O_3 crucible for 8.5 hours at 925° C, and air quenched to room temperature. The samples were then reground, pressed into disks at 200 MPa and sintered in oxygen at 960° C for 10 hours, and finally slow cooled in O_2 over a period of about 12 hours. The final step of the cooling was a 5 hour anneal in O_2 at 400° C. To produce the insulating analogue the superconducting pellet was cut into two halves, each about 0.7 grams in mass. One of the halves was annealed in vacuum at 500° C to decrease the oxygen concentration. The nominal composition of this sample was YBa$_2$Cu$_3$O$_{5.6}$, assuming that mass change corresponded entirely to oxygen loss.

The superconducting sample was single phase YBa$_2$Cu$_3$O$_{6+\delta}$ having a grain size of 20-40 μm and a bulk density of about 4.9 g/cm^3. Inductive measurements showed a transition to the superconducting state at 90 Kelvin, with a transition width of less than two Kelvin. The vacuum annealed sample was insulating at room temperature and showed no Meissner effect down to 77 Kelvin.

The samples were measured using a quasiadiabatic method. The temperature was measured with a calibrated bismuth ruthenate thin film thermometer.[6] Heat was supplied by a 1000 Ω strain gauge resistor[7] mounted with a minimum amount of GE 7031 varnish on the opposite end of the sample pill from the thermometer. The heat capacity of the addenda was calculated to be no more than 10% of that of the sample. For each data point the sample was allowed to thermally equilibrate to the regulated heat sink, then heated for a duration of one to eight seconds, a time much shorter than the thermal relaxation time through the electrical leads. The heat capacity was determined from the magnitude of the temperature excursion. The plots of heating and thermal relaxation indicated good thermal contact between the heater, sample, and thermometer. The measurement technique was checked by measuring the specific heat of a high purity copper sample.

In Figure 1 we display the specific heat of the superconducting and insulating pills, in the format C/T versus T^2. The C/T is approximately linear at higher T^2 as expected from the Debye model of the lattice heat capacity. Both samples have a slope of $2.13 \pm 0.7 \times 10^{-6}$ J / gm-K^4 (or $\Theta_D = 262$ Kelvin) indicating that the difference in oxygen concentration has little effect on the phonons. At the low T^2 end C/T rises with decreasing temperature. This behavior is very similar to that produced by small amounts of magnetic impurities in other materials.[8] In another sample we have measured which used higher purity starting materials, the low temperature "anomaly" was significantly smaller in magnitude. In any event the vacuum annealing had no effect upon this contribution.

The primary difference in the specific heats of the superconducting and the insulating samples was the magnitude of the linear contribution to the specific heat which was found to be 22 μJ / gm-K^2 in the oxygenated sample and 8 μJ / gm-K^2 in the vacuum annealed sample. Assuming a molecular weight of 666 grams, the difference in γ between the conducting and non-conducting samples was 14

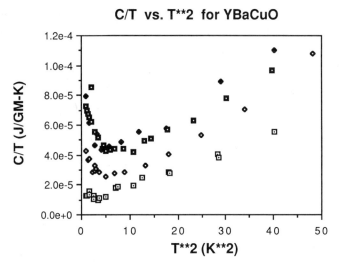

Fig. 1 The specific heat for samples of $YBa_2Cu_3O_{6+\delta}$ plotted as C/T vs. T^2. ■ Superconducting (oxygenated) and ◆ non-conducting (vacuum annealed) samples. ◆ Superconducting sample produced by same process as above but from an earlier batch. ▣ Multiphase YBaCuO sample produced at ORNL (Ref. 5) using higher purity starting materials.

$\mu J/gm-K^2$ or 9.3 mJ/mol-K^2, which is comparable to that found in the 40 Kelvin family of compounds.

Since the vacuum annealed sample had a significant linear specific heat contribution, this contribution must have a source not in conducting electrons but in some other phenomena. It has long been known that glassy materials have an excess contribution to the specific heat at low temperatures that is also linearly proportional to the temperature. In these materials the term γ has values in the range of .5 to 65 $\mu J/gm-K^2$. The linear specific heat term for our vacuum annealed sample would then fall within that range.

But in what sense could the sintered mixed oxide sample be a glass? We propose two possibilities. First, the sample is essentially a collection of small grains, separated by off-stochiometry or glassy phases at the boundaries. It has been shown that packed powders of insulators also have a glass like linear specific heat due to the grain size setting a cutoff on the phonon wavelengths.[9] A second possibility is that there is quantum tunnelling of the oxygen atoms to vacant sites in the lattice. To our knowledge the energies of the barriers to tunneling have not been calculated. The size and coupling of the grains as well as the concentration of oxygen vacancies are properties of the particular type of processing of the superconductor. The superconducting transition temperature is, within limits, not particularly sensitive to either of these properties. Therefore to determine the specific cause of the excess linear specific heat would require more sensitive characterizations of grain properties and oxygen concentrations by microscopic determination methods.

In summary, we have measured the low temperature specific heat of superconducting and non-conducting samples of the $YBa_2Cu_3O_{6+\delta}$ compound, and in each have found a contribution linearly proportional to the temperature. From the difference between of these samples we find that the linear specific heat presumably associated with the normal electrons is comparable to that found in the 40 Kelvin superconductors. There is an additional linear contribution which probably arises from glass-like excitations in the solids.

ACNOWLEDGMENTS

We are grateful to R. G. Goodrich for several helpful conversations. This research was partially supported by a contract with the Naval Air Systems. Acknowledgement is made to the Donors of the Petroleum Research Fund, administered by the American Chemical Society for partial support of this research.

REFERENCES

1. M. E. Reeves, T. A. Friedmann and D. M. Ginsberg, Phys. Rev. B35, 7207 (1987).
2. B. D. Dunlap, M. V. Nevitt, M. Slaski, T. E. Klippert, Z. Sungaila, A. G. McKale, D. W. Capone, R. B. Poeppel, and B. K. Flandermeyer, Phys. Rev. B35, 7210 (1987).
3. L. E. Wenger, J. T. Chen, Gary W. Hunter, E. M. Logothetis, Phys. Rev. B35, 7213 (1987).
4. E. Zirngiebl, J. O. Willis, J. D. Thompson, C. Y. Huang, J. L. Smith, Z. Fisk, P. H. Hor, R. L. Meng, C. W. Chu, and M. K. Wu, Solid State Commun. 63, 721 (1987).

5. Angus I. Kingon, S. Chevacharoenkul, John Mansfield, Jorulf Brynestad and David G. Haase, Advanced Ceramic Materials 2, 678 (1987).

6. Q. Li, C. H. Watson, R. Goodrich, D. G. Haase and H. Lukefahr, Cryogenics, 26, 467 (1986), and M. S. Love and A. C. Anderson, Rev. Sci. Instrum. 58, 1113 (1987).

7. D. Moy and A. C. Anderson, Cryogenics 23, 330 (1983).

8. R. O. Pohl, in Amorphous Solids: Low Temperature Properties, ed. by W. A. Phillips (Springer-Verlag, New York, 1981) p. 44.

9. R. H. Tait: Ph.D. Thesis, Cornell University (1975), as referred to in Ref. 8.

26

Processing and Superconducting Properties of $GdBa_2Cu_3O_{7-z}$

CHINNARONG ASAVAROENGCHAI, ROBERT M. METZGER, DAVID A. ROBINSON, ANNY MORROBEL-SOSA, and JOSEPH S. THRASHER Department of Chemistry, University of Alabama, Tuscaloosa AL 35487-9671

CHESTER ALEXANDER, Jr. Department of Physics, University of Alabama Tuscaloosa AL 35487-9671

DONALD A. STANLEY and M. ABBOT MAGINNIS Tuscaloosa Research Center, U. S. Bureau of Mines, Tuscaloosa AL 35486-9777

1. INTRODUCTION

In April 1986 Bednorz and Mueller (1) reported superconductivity in the quaternary mixed-valence system $La_xBa_{2-x}CuO_{4-z}$, with onset temperatures in excess of 30 K, several degrees higher than the previous record-holder, Nb_3Ge (2). In January 1987 the Houston-Alabama (Huntsville) group of Chu and Wu found that black, single-phase $YBa_2Cu_3O_{7-z}$ superconducts with onset at 93 K and T_c ("zero" resistance) = 80 K (3). Within a few months it was found in laboratories around the world that there were several such compounds with T_c in the 90-100 K range, i.e. well above the boiling temperature of liquid nitrogen (4-6).

This superconductivity is observed in compounds $R_xA_yCuO_w$ in which a certain amount (x) of trivalent rare earth ion (R) (or its analog) coexists with a divalent alkaline earth ion (A) and also a certain fraction of the oxide sites in the lattice are vacant; all this makes possible a mixed-valence Cu(II)-Cu(III) state, in which the copper (a Jahn-Teller ion) can be either square-planar coordinated or square-pyramidally coordinated in a one- or two-dimensional array.

So far, two families of crystal structures exhibit superconductivity. The first family, with T_c around 40-50 K, is typified by $La_{2-x}Ba_xCuO_{4-z}$, which has <u>nominal</u> Cu oxidation number CON = 2+x-2z; the second family has T_c = 90-100 K, and is typified by $YBa_2Cu_3O_{7-z}$, with CON = 2.333-0.666z. In both families, there are planar or almost planar ("dimpled") arrays of Cu and O atoms, of stoichiometry CuO_2, surmounted by apical O atoms, yielding a square-pyramidal (i.e. distorted octahedral) coordination about Cu; these are the "2-D planes", possibly responsible for T_c around 40 K. In the $YBa_2Cu_3O_{7-z}$ structure there is also a ribbon motif of square-planar Cu and O atoms, of stoichiometry CuO_3; this "1-D ribbon" may be responsible for T_c around 90 K (7).

We first achieved superconductivity in a rapid-quenched green-black composite $Gd_3Ba_3Cu_4O_z$, with T_c = 62 K (onset at 71 K) (8). We have also found superconductivity in single-phase $YBa_2Cu_3O_{7-z}$, with T_c = 90 K (onset at 96 K) (9). Here we report on the synthesis, powder X-ray diffraction, and conductivity of $GdBa_2Cu_3O_x$, in which we find T_c = 100 K (onset at 108 K).

2. EXPERIMENTAL

The $GdBa_2Cu_3O_x$ samples used for this study were prepared from appropriate molar mixtures of fine powders of Gd_2O_3 (Aesar), $BaCO_3$ (Mallinckrodt), and CuO (Aldrich), all with purity > 95%, ground together in a ceramic mortar and pestle. Pressed pellets (1.2 kbar), placed in a ceramic boat, were sintered in a Lindberg tube furnace at 950°C for 12 hrs. in flowing oxygen. This step was repeated for 16 hrs. after a regrinding and pressing of the samples. The pellets were then annealed at 500°C for 5 hrs. Initial dc electrical resistance measurements did not indicate the presence of a superconducting transition; the samples were therefore submitted to further sintering in flowing oxygen at 980°C for 18 hrs. All measurements reported here were done on the resulting black samples. A small specimen was studied by X-ray diffraction (Philips 3100 generator).

Measurements of ac and dc electrical resistance versus temperature were made on samples that were attached to the cold finger of a closed

Table 1.

X-ray powder diffractogram for GdBa$_2$Cu$_3$O$_{7-z}$ [λ (Cu Kα$_1$) = 1.540598 Å,

λ(Cu Kα$_{ave}$) = 1.541877 Å], was indexed to yield **a** = 3.855(5) Å, **b** = 3.872(4)

Å , **c** = 11.348(11) Å, in agreement with the values **a** = 3.89 Å, **b** = 3.89 Å, **c** =

11.73 Å of Ref. (10).

d_{obs}/Å	I/I_o	h k l	d_{obs}/Å	I/I_o	h k l	d_{obs}/Å	I/I_o	h k l
1.508	3	0 0 1	2.550	1	(b)	1.631	3	0 0 7
3.862	28	1 0 0	2.489	7	1 1 2	1.582	11	1 2 3
3.544	1	——	2.329	21	(c)	1.571	3	(d)
3.136	1	(a)	2.228	4	1 1 3	1.529	1	——
2.907	1	——	1.939	100	2 0 0	1.480	3	2 1 4
2.833	4	——	1.913	5	2 0 1	1.456	<1	——
2.733	12	1 1 0	1.733	2	1 2 0	1.423	2	0 0 8
2.716	20	1 0 3	1.665	14	——	1.382	2	2 1 5
2.616	2	——	1.664	9	0 0 8	1.365	5	2 2 1

(a) attributed to the 222 reflection   of Gd$_2$O$_3$ (**a** = 10.813 Å) , or to the 111
    reflection of BaO (**a** = 5.523 Å)
(b) attributed to the 111 reflection   of CuO (**a** = 4.684 Å, **b** = 3.425 Å, **c** =
    5.120 Å, β =99.47°)
(c) attributed to the 11-1 reflection of CuO
(d) attributed to the 2 0 2 reflection   of CuO

cycle refrigerator unit (CTI-Cryogenics 22C). Temperature was measured
with a chromel versus Au-0.07% Fe thermocouple attached to the sample
with electrical insulating varnish (GE 7031); the data were digitized with
a Scientific Instruments temperature indicator. Four parallel leads
were painted on the sample with conductive silver paint (DuPont 7713).
The small amount of solvent present in the paint is evaporated by
heating the pellets to 540°C in a flow of oxygen for approximately 10
min. Then copper wire contacts were soldered onto the painted strips.
Data acquisition and analysis were performed with a personal computer
(HP 9816) and a data acquisition/control unit (HP 3497A).

The four-probe dc measurements were made by maintaining a constant sample current of less than 1.5 mA, and reading the voltage drop across the inner leads on the sample using a nanovolt amplifier (Keithley 140). The four-probe ac measurements were performed using, as the constant current source, the 2 kHz, 0.747 V reference output voltage of a lock-in amplifier (PAR 120) through a 1 kΩ resistor in series with the sample. The sample voltage was pre-amplified and detected by the signal channel of the lock-in.

3. RESULTS AND DISCUSSION

The powder diffractogram consists of 27 lines, of which 17 belong to a unit cell of dimensions very similar to those of $YBa_2Cu_3O_{7-z}$, and also very similar to that reported elsewhere for $GdBa_2Cu_3O_x$ (10). Of the 10 other lines, 2 or 3 can be attributed to CuO; the rest are very weak.

The plot of the dc and ac resistance versus temperature is given in Figs. 1 and 2 respectively. The onset temperature for superconductivity was 110 K for both measurements. T_c, the temperature where "zero" resistance was reached was 102 K (dc, Fig. 1) and 101 K (ac, Fig. 2).

The superconducting critical temperature reported here [$T_c = 100$ K (onset at 108 K)] is a few degrees higher than those reported by other groups. Hor et al. (10) measured $T_c = 92$ K (onset at 95 K); Mei et al. (11) found $T_c = 87$ K (onset at 95 K); Fisk et $al.$ (12) found $T_c = 95 \pm 3$ K, and, later, $T_c = 95 \pm 1$ K (13). We are confident about our thermometry, and feel that the X-ray data show evidence of a mostly pure phase. However, the presence of a few reflections which can be attributed to CuO may indicate some non-stoichiometry in our preparation which may, in turn, assist in the modest increase in transition temperatures. In fact, similar, if broad, transitions at relatively high temperatures were reported earlier for samples of yttrium barium copper oxides with deviations from the "1-2-3" stoichiometry (14).

Preliminary Hall effect studies were performed by measuring the resisitivity of the sample in an applied magnetic field of 0.3 T.

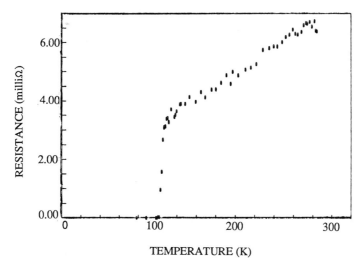

FIGURE 1. Resistance (dc, mΩ) of a pellet of GdBa$_2$Cu$_3$O$_{7-z}$ as a function of temperature (degrees K).

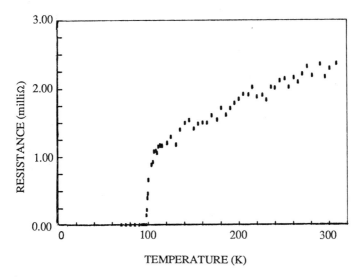

FIGURE 2. Resistance (ac, mΩ) of a pellet of GdBa$_2$Cu$_3$O$_{7-z}$ as a function of temperature (degrees K).

ACKNOWLEDGMENTS

We are grateful to Prof. J. H. Fang of the Department of Geology for providing the X-ray data, and to Prof. Geoffrey N. Coleman for supplying large amounts of assorted rare earths.

REFERENCES

1. J. G. Bednorz and K. A. Mueller, Z. Phys. B **64**, 189 (1986).

2. J. R. Gavaler, Appl. Phys. Lett. **23**, 480 (1973).

3. M. K. Wu, J. R. Ashburn, C. J. Torng, P. H. Hor, R. L. Meng, L. Gao, Z. J. Huang, Y. Q. Wang, and C. W. Chu, Phys. Rev. Lett. **58**, 908 (1987).

4. R. J. Cava, B. Batlogg, R. B. van Dover, D. W. Murphy, S. Sunshine, T. Siegrist, J. P. Remika, E. A. Rietman, S. Zahurak, and G. P. Espinosa, Phys Rev. Lett. **58**, 1676 (1987).

5. A. M. Stacey, J. V. Badding, M. J. Geselbracht, W. K. Ham, G. F. Holland, R. L Hoskins, S.W. Keller, C. F. Millikan, and H.-C. zur Loye, J. Am. Chem. Soc. **108**, 2528 (1987).

6. H. Takagi, S.-I. Uchida, K. Kishio, K. Kitazawa, K. Fueki, and S. Tanaka, Jpn. J. Appl. Phys. **26**, L320 (1987).

7. M. H. Whangbo, M. Evain, M. A. Beno, and J. M. Williams, Inorg. Chem., **26**, 1831 (1987).

8. D. A. Robinson, A. Morrobel-Sosa, C. Alexander, Jr., J. S. Thrasher, C. Asavaroengchai, and R.M. Metzger, in Proceedings of the NATO Workshop on Inorganic and Organic Low-Dimensional Crystalline Materials (Minorca, 3-8 May 1987), P. Delhaes, Ed., Plenum, in press.

9. D. A. Robinson, A. Morrobel-Sosa, C.Alexander, J. S. Thrasher, C. Asavaraoengchai, R. M. Metzger, D. A. Stanley, and M. A. Maginnis, J. Chem. Phys., submitted.

10. P. H. Hor, R. L. Meng, Y. Q. Wang, L. Gao, Z. J. Huang, J. Bechtold, K. Forster, and C. W. Chu, Phys. Rev. Letters **58**, 1891 (1987).

11. Y. Mei, S. M. Green, G. G. Reynolds, T. Wiczynski, H. L. Luo, and C. Politis, Z. Phys. B **67**, 303 (1987).

12. Z. Fisk, J. O. Willis, J. D. Thompson, S.- W. Cheong, J. L. Smith, and E. Zirngiebl, J. Magnetism and Magn. Materials, submitted.

13. S. E. Brown, J. D. Thompson, J. O. Willis, R. M. Aikin, E. Zirngiebl, J. L. Smith, Z. Fisk, and R. B. Schwartz, Phys. Rev. B **36**, 2298 (1987).

14. C. Politis, J. Geerk, M. Dietrich, B. Obst, and H. L. Luo, Z. Phys. B **66**, 279 (1987).

27

Processing and Microstructures of $YBa_2Cu_3O_{7-\sigma}$

ANGUS I. KINGON, SOPA CHEVACHAROENKUL, STANE PEJOVNIK,[*] RICARDO VELASQUEZ, RICHARD L. PORTER, THOMAS M. HARE, and HAYNE PALMOUR III   Department of Materials Science and Engineering, North Carolina State University, Raleigh, North Carolina 27695-7907

DAVID G. HAASE   Department of Physics, North Carolina State University, Raleigh, North Carolina 27695-8202

1.   INTRODUCTION

The high T_c superconducting ceramic $YBa_2Cu_3O_{7-\sigma}$ has been prepared by a large number of groups around the world. As expected, the reported values of the primary superconducting properties such as T_c, which are dependent upon crystal and electronic structure, do not vary a great deal, as long as the oxygen content is correctly controlled. However, an analysis of recently published results (1-4), coupled with our own research results, confirm that a great deal of progress must still be made towards controlling the tertiary superconducting properties such as critical current density, J_c, and AC losses, which are determined primarily by microstructure. In addition, the microstructure (grain size, porosity, grain boundary phases) has a major influence on mechanical properties as well as stability under given environmental conditions. At the present time the densities of polycrystalline ceramics are low, as are J_c values, and microstructures are clearly far from optimum. The majority of workers (1-4) report densities < 90% of theoretical.

In this paper we address aspects of three major issues in the processing of $YBa_2Cu_3O_{7-\sigma}$:
   a.   The mixed oxide reaction to form the compound.
   b.   The formation of liquid phases during sintering and the control of densification and microstructure.

---

*On leave from the Institute of Chemistry *"Boris Kidric"*, Ljubljana, Yugoslavia

335

c. The ordering of O-vacancies, and the effect on properties.

These topics are discussed in turn in the following sections.

2.   EXPERIMENTAL

The raw materials used for this study, unless otherwise stated, were $BaCO_3$ (Fisher, 99.7%), CuO (Alfa Products, ACS grade), and Y_2O_3 (Aesar, 99.9%).

2.1. Mixed-Oxide Reaction Study

Reagents were extensively mixed by hand, and then for 15 minutes in a Spex mill (Spex Industries, Inc., Metchen, N.J.). Samples were placed in Al_2O_3 crucibles, and reacted (calcined) isothermally for a given time in a muffle furnace with an air atmosphere. The reaction products were characterized by X-ray diffraction. The extent of reaction was monitored semi-quantitatively by comparison of the heights of the (103)+(110) peak of $YBa_2Cu_3O_{7-\sigma}$, and the (100) peak of $BaCO_3$. Calibration was performed using $BaCO_3$ and $YBa_2Cu_3CuO_{7-\sigma}$ standards.

2.2   Sintering Studies

Typical conditions were as follows:

Raw materials were mixed as described above, or mixed in a vibratory mill. Mixed powders were loosely pressed into a disk, and reacted in Al_2O_3 crucibles for 12 hours at 890°C in air. The disks were ground in a mortar and pestle, repressed and reacted again for 12 hours at 930°C. The reaction procedure was repeated a third time, 12 hours at 900°C. The samples were finally milled in a vibratory mill using isopropanol, and zirconia milling media. Some variations were introduced into the procedure in order to investigate the effect of specific processing steps. Disks were pressed uniaxially or isostatically at 250 MPa. The sintering studies utilized a precision digital dilatometer which has been described previously (5). Unless otherwise noted, the dilatometric sintering experiments were carried out in a controlled flow of O_2.

2.3   Ordering of Oxygen Vacancies

Powder compacts were prepared as described above, sintered in a tube furnace at 960°C for 10 hours, then quenched to room temperature in < 3 minutes in flowing

O_2, or alternatively slow cooled in O_2, including an O_2 anneal at 450°C for 6 hours. It should be noted that these samples were ~ 80% of theoretical density, and contained *open* porosity.

The samples were milled to electron transparency in a Gatan® dual-beam ion mill at 6 kV, 15° and 1 mA until a small perforation occurred. They were further milled at 5 kV, 10° and 1 mA for another 30 minutes. The TEM studies were performed in a Hitachi H-800 and a Phillips EM 430T.

3.   RESULTS AND DISCUSSION

3.1   Mixed-Oxide Reaction Study

The X-ray diffraction results for the reaction study are presented in Table 1. It is clear that the reaction proceeds at reasonable rates only at temperatures greater than 850°C in these mixed oxide samples. Cation interdiffusion distances would have to be significantly reduced in order to lower the reaction temperatures. Differential Thermal Analysis (DTA) also indicates that the increase in the rate above 850°C coincides with the formation of liquid phases at $\geq$ 890°C (see 3.2).

Table 1

Extent of Reaction Determined by Comparison of X-Ray Diffraction Intensities
$YBa_2Cu_3O_{7-\sigma}I_{103,110}/BaCO_3I_{100}$

Reaction Temperature (°C)	Mass Isothermal Reaction time Ratio $YBa_2Cu_3O_{7-\sigma}$: $BaCO_3$		
	22 mins	60 mins	180 mins
900		90:10	
850	29:1	42:58	66:34
800		22:78	
750		7:93	11:89

Several observations are of interest. Firstly, there was no lower temperature regime in which binary compounds were formed, as is common for the formation of oxides with three or more components (6). Only at 900°C were traces of Y_2BaCuO_5 and $BaCuO_2$ observed. All further diffraction peaks were accounted for, with the exception of one observed at the intermediate temperatures.

Secondly, it is of interest that the $YBa_2Cu_3O_{7-\sigma}$ forms with the A-site cations ordered within the lattice, *i.e.*, there is no observable temperature or time regime where partial disordering of the Y^{3+} and Ba^{2+} ions is observed. This can be deduced from the shape and relative intensity of the (113) peak at 40.39 °2θ (CuK_α radiation).

Thirdly, we have observed that while powder X-ray diffraction indicates that the reaction has gone to completion, TEM and DTA investigations *always* show the presence of some of unreacted precursors, or other compounds, typically Y_2BaCuO_5 or grain boundary glassy phases (7). This situation becomes worse in the case of higher purity $BaCO_3$ (99.999%) with evidence of $BaCO_3$ remaining despite five reactions. Unreacted $BaCO_3$ is easily identifiable using a DTA by the enthalpy associated with the solid state γ to β transition at ~ 811°C (heating).

Table 2

Relation between processing route and fired density

Process Route Investigated	Fired density $[\rho, g/cm^3]$	Percent of ρ_{th}*
Single Calcined, Spex Milled	4.75	~ 75%
Double Calcined, Spex Milled	5.23	~ 83%
Double Calcined, Vibratory Milled	5.77	~ 92%
Double Calcined, Dilatometer Firing	6.01	~ 95%

*$\rho_{th} = 6.305$g cm^{-3}

3.2   Sintering Studies

The sintered microstructure is extremely process sensitive. Selected data are presented in Table 2 to illustrate this point.

Parameters such as particle size and size distribution have a major influence, but are not discussed in this paper. Instead we focus on the influence of liquid phases on the microstructural evolution.

The presence of liquid phases can be deduced by DTA, from the dilatometer densification data, and by direct observation using a scanning electron microscope (SEM). DTA analysis of unreacted mixed powders show the γ to β transition of $BaCO_3$ at ~811°C, followed by an endothermic reaction or melting beginning at ~960°C. Further cycling, or a single reaction, results in one or more liquid phases, the observed onset temperature as low as ~830°C, and moving up in temperature to ~920°C with increased time at temperature, or repeated cyclings. The onset of incongruent melting was then observed at about 980°C.

Repeated reaction eliminated the liquid phases below ~930°C, but even after five reactions, the heat effect associated with another melting transition was always observed, with an onset in the range 930-960°C, followed by incongruent melting. As stated previously, the postulate that this is a liquid phases was confirmed by dilatometry and by SEM of fractured and polished samples. The presence of a liquid phase is consistent with conclusions of a low temperature specific heat study on these materials (9).

The liquid phases have a major influence on the microstructural evolution. Fig. 1. shows the shrinkage behavior of a single calcined (reacted) sample, as well as those of that of double and triple calcined samples. This and other data indicate that the initial liquid phases formed, in the temperature range 830-930°C, have a deleterious effect on microstructure, sometimes resulting in sample expansion rather than shrinkage. This may be understood in terms of the competing diffusion processes for reaction and densification. In the case of the singly calcined sample shown in Fig. 1 (a), no densification was observed up to 960°C (data shown up to 920°C only).

The liquid phase which has an onset of melting in the range 930-960°C is more useful for densification (Fig. 1c), as long as the conditions are well controlled.

The kinetics of the resulting densification is typical of liquid phase sintering, occurring in a narrow temperature regime. As frequently occurs for liquid phase sintering, the densification process is rate dependent, and this is dramatically illustrated in Table 3.

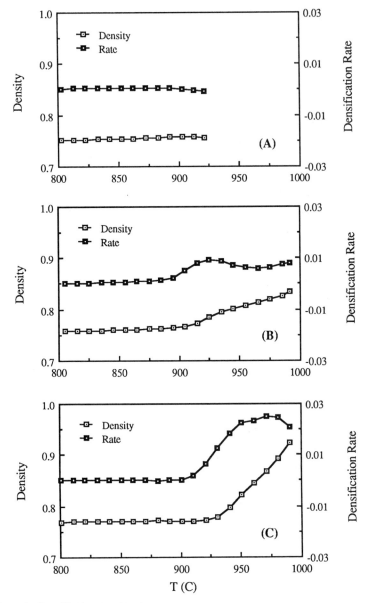

Figure 1. Densification data (density and densification rate) for $YBa_2Cu_3O_{7-\sigma}$ at a heating rate of 10°C min.[-1]. (A) Single calcination. (B) Double calcination. (C) Triple calcination

Table 3

Effect of heating rate on sintered density (The heating rate dependence is due to liquid phase formation*)

Heating Rate From 800°C to 960°C [°C/min]	Density [g/cm^3]	% of Theoretical Density**	Mean grain size (μm)$^\Delta$
10	5.55	88	4
2.5	5.86	93	7
1	5.86	93	10[†]
0.25	6.05	96	19

*Heating rate to 800°C - 10°C/min
**Theoretical density = 6.305g cm^{-3}
†Anisotropic microstructure - elongated grains
ΔEstimated by linear counting on fracture surfaces.

The following points are pertinent:

a. The reason for the liquid phase at 930-960°C is not yet fully understood. *"Pure"* single phase $YBa_2Cu_3O_{7-\sigma}$ melts incongruently at ~ 1010°C (8), but slight departures from *"exact"* stoichiometry can result in liquid phases, with eutectic temperatures in the range 930-960°C (8). We therefore deduce that the mixed oxide reaction results in a compositional distribution which is difficult to homogenize, and is not eliminated even with repeated calcining under the conditions selected.

b. Identification and characterization of the liquid phases is currently being investigated by microanalysis. Important data such as wetting angles do not yet exist. Off-stoichiometric compositions are also intentionally being prepared. The effect of a small copper excess or deficiency is illustrated in Fig. 2, showing a major effect on the microstructure. We deduce that this is directly related to the different compositions of second phases in accordance with the phase equilibria (8). Details will be published separately (10).

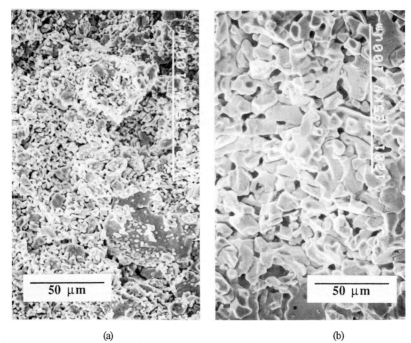

(a) (b)

Fig. 2:   Comparison of microstructures of samples with a small Cu excess or deficiency. (a)
YBa$_2$Cu$_{2.91}$O$_{7-\sigma}$ (b) YBa$_2$Cu$_{3.09}$O$_{7-\sigma}$ (overall compositions).

  c. The effect of the liquid phases or off-stoichiometric compositions on
tertiary properties such as J_c have not yet been determined.

  d. Under the optimum processing conditions, dense samples have been
obtained, containing closed rather than open porosity. This has an effect
on O-transport, as discussed in the following section.

3.3 Ordering of Oxygen Vacancies

It is well known that control of the oxygen stoichiometry is required in order to
obtain optimum superconducting properties. It has generally been believed that after
sintering it is essential to cool extremely slowly in O$_2$, or to anneal in O$_2$ at 400-
500°C, in order to obtain sharp transitions. Furthermore, Beyers *et al* (10)
concluded that it is essential to order the oxygen vacancies in the basal plane, *i.e.*,
the CuO$_3$ chains. Our results described below indicate that this is an incomplete
picture. A series of samples were cooled from the sintering temperature under a
variety of conditions, as shown in table 4.

Table 4

Treatment[†]	Oxygen Content	Superconductivity
O_2, slow cool, 500°C anneal	6.9*	92K onset, sharp
O_2 quench to R.T.	6.82	92K onset, sharp
Air, furnace cool	6.81	87K onset (ΔT_c = 10K)
Air, quench**	6.69	Usually semiconductor behavior (see Fig. 5)

* This value is the assumed reference. All other values by difference relative to slow cooled in O_2.

** O content and properties usually more variable from run-to-run in the case of samples quenched in air.

† The samples were ~ 80% of theoretical density, and contained opened porosity.

For these samples, which contained open porosity, it is clearly possible to cool extremely rapidly to room temperature, if an oxygen atmosphere is maintained. The susceptibility data for these two samples are shown in Fig. 3.

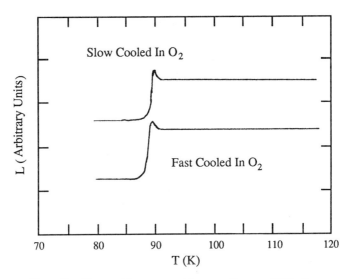

Figure 3.   The transition, measured by magnetic susceptibility for different cooling conditions in O_2.

Of interest is the subgrain microstructures of the O_2 quenched samples, in comparison with the slow-cooled and annealed samples. The width of the twins viewed along [001] in slow cooled samples are typically 100-200 nm wide (Fig. 4 (a)). However, in the fast cooled samples, the size of the twins are considerably reduced. Fig. 4 (b) shows a grain of the fast cooled sample showing twins of smaller dimension. Orthogonal boundaries are also observed, indicating the presence of both 110 and $1\bar{1}0$ twin boundaries. Fig. 4 (c) shows twins of larger dimension in the center of the grain, but the hatched region toward the outer boundaries of the grain indicate both 110 and $1\bar{1}0$ twin boundaries, with substantially smaller domain dimensions. The implication is that the range of ordering of the O vacancies in this region is greatly reduced, in the order of several nanometers. The gradation of the size of the twin region can be rationalized in terms of the longer times required for oxygen to diffuse to the center of the grains, and thus the longer times available to achieve ordered vacancies.

It is important to note that in contrast with the above porous samples, dense samples with only closed porosity ($\geq 94\%$ of theoretical density) were significantly more difficult to anneal to achieve the correct oxygen stoichiometry, and thus desired properties. These samples required O_2 annealing times at 500°C of > 48 hours in order to achieve a significant Meissner effect. Using the data above, we can calculate that lattice diffusion is *not* the limiting step. We therefore deduce that the grain boundary phase(s) are impermeable to oxygen. We are presently studying this oxygen equilibration using samples which are purposefully off-stoichiometry (11).

4. CONCLUSIONS

a. The mixed oxide reaction proceeds at a significant rate only above 850°C. No temperature/time regime was identified in which binary compounds were formed, rather than the ternary compound $YBa_2Cu_3O_{7-\sigma}$. Furthermore, the compound forms *without* a temperature or time regime where the A-site cations are disordered. TEM and DTA results indicate that low concentrations of second phases remain even after repeated reactions.

b. The density and microstructure of sintered $YBa_2Cu_3O_{7-\sigma}$ is extremely process sensitive. Liquid phases dominate the microstructural evolution. These liquid phases have characteristic enthalpies of melting. Incomplete reaction results in at least one liquid phase melting in the range 820-930°C.

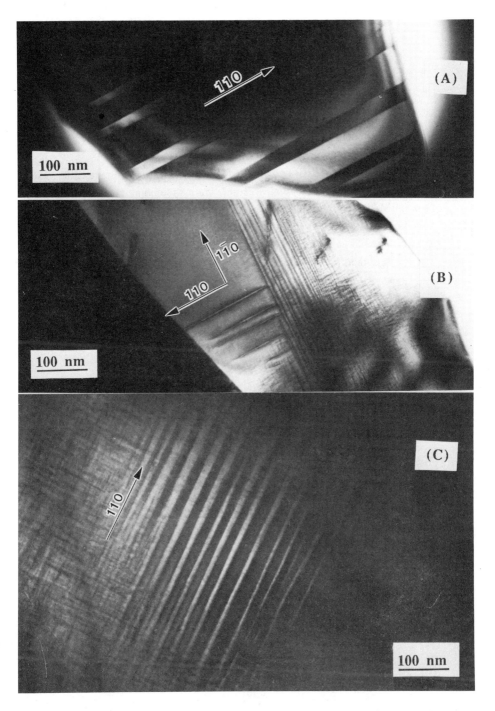

Figure 4. Subgrain microstructure showing twins. (A) grain typical of an O_2 slow cooled or annealed sample. (B) and (C) Grains typical of porous samples quenched in O_2.

A further liquid phase observed in all samples, has an onset of melting in the range 930-960°C, and this phase typically results in liquid phase densification. The reason for this liquid phase is not yet clear, as "pure" $YBa_2Cu_3O_{7-\sigma}$ should melt incongruently at ~ 1010°C. Proper processing results in dense (95% of theoretical) $YBa_2Cu_3O_{7-\sigma}$. This density has been achieved by only a few research groups.

c. Porous samples of $YBa_2Cu_3O_{7-\sigma}$ can be quenched to room temperature in an O_2 atmosphere, and still retain a sharp superconducting transition. TEM studies show the twin structure to be extremely small, containing both (110) and ($1\bar{1}0$) twin boundaries. This implies that the range of ordering of O-vacancies and CuO_3 chains is small, and this ordering is less critical than previously believed. With dense samples it is difficult to achieve the required oxygen concentration.

ACKNOWLEDGEMENTS

We acknowledge contributions made by Mr. Roger Russell and Mr. David Chapman. Valuable discussions were held with Drs. Carl Koch, Bob Davis, Don Kroeger (ORNL), Jorulf Brynestad (ORNL), John Mansfield (U. of Michigan) and Mike Mullins (RTI). Work was partially performed under a grant from the Alternate Energy Corporation of North Carolina.

REFERENCES

1. Papers in the Jap. J. of Appl. Phys., 26 [4] L311-L525.
2. Extended astracts published in the Proceedings of Symposium S, "High Temperature Superconductors", 1987 Spring Meeting of the Materials Research Society, April 23-24, Anaheim California. Eds., S. U. Gubser and M. Schlutev. Published by the Materials Research Society.
3. Papers in Advanced Ceramic Materials, 2 [3B] 1987, especially Section II: Processing and Fabrication.
4. Papers in The Chemistry of High-Temperature Superconductors, ACS Symposium Series no. 351, Eds., D. L. Nelson, M. S. Whittingham and F. George (1987).
5. a. A. D. Batchelor, M. J. Paisley, T. M. Hare and H. Palmour, III, Precision Digital Dilatometry: a microcomputer-based approach to sintering studies, pp. 233-251, Eds., R. F. Davis, H. Palmour, III and T. M. Hare, Processing of Crystalline Ceramics, Mat. Sci. Research, Vol. 22, Plenum Press, New York, 1978.

   b. H. Palmour, III and T. M. Hare, Rate Controlled Sintering Revisited, pp. 17-34, Eds., G. C. Kuczynski, D. P. Uskokovic', H. Palmour, III and M. M. Ristic', Sintering '85. Plenum Press, New York-London, 1978.

6. A. I. Kingon, B. V. Hiremath and J. V. Biggers, J. Am. Ceram Soc., 66 [11], 790-793 (1983).

7. A. I. Kingon, S. Chevacharoenkul, J. Mansfield, J. Brynestad and D. G. Haase in Reference 3, 678-687.

8. D. G. Haase, R. Velasquez and A. I. Kingon, The low temperature specific heat of YBa$_2$Cu$_3$O$_{6+\sigma}$ compounds", submitted to the Proceedings of the Conference on High T$_c$ Superconductors, Chapel Hill, North Carolina, September 1987.

9. For example, R. S. Roth, K. L. Davis and J. R. Dennis in Reference 3, 303-312.

10. R. Beyers, G. Lim, E. M. enger, V. Y. Lee, M. L. Ramirez, R. J. Savoy, R. D. Jacowitz, T. M. Shaw, S. Laplaca, R. Boéhme, C. C. Tsuei, Sung I. Park, M. W. Shafer, W. J. Gallagher and G. V. Chandrashekhas, submitted to Appl. Phys. Lett.

11. S. Pejovnik, T. M. Hare, A. I. Kingon, R. L. Porter and H. Palmour, III, Sintering and Microstructure Development in Superconducting YBa$_2$Cu$_3$O$_{7-\sigma}$, Invited Paper to be presented in Sintering '87, Tokyo, Japan, November 4-6, 1987.

28

Raman Spectroscopic Characterization of Different Phases in the Y-Ba-Cu-O System

B. H. LOO, M. K. WU, D. H. BURNS, A. IBRAHIM, C. JENKINS, T. ROLIN, and Y. G. LEE   Departments of Chemistry and Physics, University of Alabama—Huntsville, Huntsville, Alabama 35899

D. O. FRAZIER   Space Science Laboratory, NASA Marshall Space Flight Center, Huntsville, Alabama 35812

F. ADAR   Instruments S. A., Edison, New Jersey 08820

1. INTRODUCTION

Subsequent to the discovery of high T_c superconductivity in the Y-Ba-Cu-O compound system by Wu et. al. (1), two phases were separated and identified (2-4): the black $YBa_2Cu_3O_{6+x}$ ("123") phase and the green Y_2BaCuO_5 ("211") phase. The black phase is found to be responsible for the high temperature superconductivity.

$YBa_2Cu_3O_{6+x}$ is usually prepared by the standard oxide-carbonate calcination method in which powdered reagents are ground together, pressed into pellets, and heated in a furnace at 950°C in the presence of oxygen. Because solid state mixing does not often achieve complete homogeneity, minority phases are frequently present in addition to the desired majority phase.

Phase equilibria in the the Y_2O_3-BaO-CuO system have been studied by three groups (5-8) whose findings are consistent. The Y_2O_3-BaO-CuO system forms a complex oxide system. The Y_2O_3-BaO binary system consists of four possible barium yttrium oxides. Only two of these, BaY_2O_4 and $Ba_4Y_2O_7$, are stable at 950°C. $Ba_2Y_2O_5$ is stable only below 950°C and $Ba_3Y_4O_9$ is stable only above 1000°C. The BaO-CuO binary system consists of barium

349

cuprate ($BaCuO_2$) and possibly dibarium cuprate (Ba_2CuO_3) and tribarium cuprate (Ba_3CuO_4). The blue-green yttrium cuprate ($Y_2Cu_2O_5$) is the most stable phase in the Y_2O_3-CuO binary system. In the Y-Ba-Cu-O ternary system several other phases such as the 211 phase and the "132" phase, $YBa_3Cu_2O_{7+x}$, are known to exist in addition to the superconducting 123 phase. Thus, under optical microscopy, superconducting materials that are nominally single phase may exhibit secondary phases whose presence can be confirmed by acquisition of Raman spectra.

2. EXPERIMENTAL METHODS

The 123 and 211 compounds were prepared by standard oxide-carbonate calcination methods. Appropriate amounts of Y_2O_3, $BaCO_3$, and CuO powder were ground together and calcined in air at $950^{\circ}C$ for 12 hours. To achieve high T_c superconductivity, the 123 compound was reground, again pressed into pellets, annealed in a flowing stream of oxygen gas at $950^{\circ}C$ for 18 hours, and followed by slow cooling in the furnace. Annealing in a vacuum at $400^{\circ}C$ for four hours converted the sample to the semiconducting tetragonal phase. Its structure was confirmed by the powder X-ray diffraction method.

Electrical resistivity measurements on the 123 compound were made using the standard four-probe technique with an alumel-chromel thermocouple used for temperature measurements. The samples exhibit linearly decreasing resistance with decreasing temperature from 300K followed by a sharp drop to the superconducting state near 90K. The superconducting onset temperature is 96K and the transition width is typically 2-3K.

Binary oxides were prepared in a similar manner. $BaCuO_2$ was prepared by grinding together equal molar quantities of high purity $BaCO_3$ and CuO, pressing the mixture into pellets and heating the pellets in a furnace at 950oC for 12 hours. $Y_2Cu_2O_5$ was prepared similarly from high purity Y_2O_3 and CuO. BaY_2O_4 and $Ba_4Y_2O_7$ were prepared by grinding together BaO, BaO_2 or $BaCO_3$ with Y_2O_3 in molar ratios of 1:1 (BaY_2O_4) and 4:1 ($Ba_4Y_2O_7$), and heating at $950^{\circ}C$ for 16 hours. Other molar ratios, 2:1, 3:1, 5:1, 6:1, were also used to check for the formation of different phases. Some pellets of binary oxides were fired several times to assure complete reaction.

Raman measurements were always made on several spots of a number of different sintered pellets of a given preparation. Samples were not polished for analysis. The instrument used at both the University of Alabama in Huntsville and at the Application Laboratory of Instruments, S. A., was a U1000/MOLE Raman microprobe using the 514.5-nm line of an argon ion laser (Spectra Physics 2020 and 164) for excitation. Power levels at the sample surface were between 1 and 10 mW. Many measurements were made with a 50x objective and with the laser beam defocused in order to acquire spectra representative of the bulk phase and/or to prevent laser damage to the sample, which occasionally occurred. The size of the sampled spot varied from about 1 to 10 microns in diameter. Spectral resolution was 3-5 cm^{-1}. All spectra reported here are reproducible.

3. RESULTS AND DISCUSSION

Our results are interpretable in terms of existing knowledge of the Y-Ba-Cu-O system (5-8). A phase diagram for the ternary oxide system, adapted from Ref. 5, is shown in Fig. 1. We will discuss our Raman results in terms of starting materials, stable phases in the three binary systems and, finally, the stable phases in the ternary system.

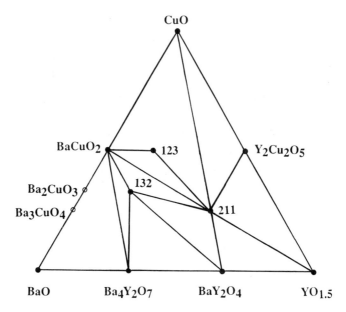

Figure 1: Phase diagram of the Y_2O_3-BaO-CuO system at 950°C.

3.1 STARTING MATERIALS

3.1.1 $BaCO_3$, BaO, BaO_2

The   Raman   spectrum   of $BaCO_3$ is shown in Fig.   2(a).   The
peaks   most useful as a fingerprint of its presence are   at   135,
154,   690,   and 1059 cm^{-1}.   The 690- and 1059-cm^{-1} bands   are   in
good agreement with those reported by Griffith (9).

Bao has a NaCl-type structure and hence does not   exhibit   first
order   Raman scattering.   The second order Raman spectrum of   BaO
(Fig. 2(b)), with several broad bands at 130, 205, 276, 340, 400,
and 650 cm^{-1}, is in good agreement with the spectrum reported   by
Rieder, et al. (10).   Care was taken during the data   acquisition
to prevent the reaction of BaO with CO_2 to form $BaCO_3$.   The BaO_2
spectrum (not shown) shows a prominent peak at 841 cm^{-1}.

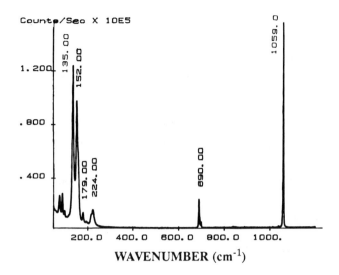

Figure 2(a): Raman Spectrum of BaCO$_3$.

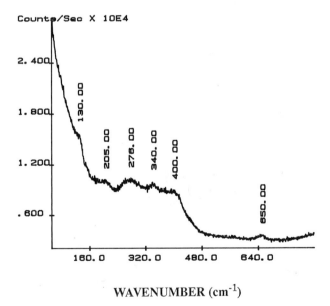

Figure 2(b): Raman Spectrum of BaO.

3.1.2 CuO

Copper (II) oxide is a very poor scatterer with weak peaks at 293, 342, and 626 cm^{-1} (Fig. 3). The peak at 321 cm^{-1} is an artifact of our optical system.

3.1.3 Y_2O_3

The spectrum of yttrium oxide is complex. Fig. 4 shows the spectrum below 1000 cm^{-1}. The most prominent peak is at 377 cm^{-1} and there are numerous weaker peaks. The 377 cm^{-1} peak was used as a diagnostic peak in monitoring the completion of reactions involving Y_2O_3 in this study.

3.2 BINARY SYSTEMS

3.2.1 $BaO-Y_2O_3$ System

Only the compounds representing the reaction of 1:1 and 4:1 molar ratios of barium oxide with yttrium oxide have been confirmed,

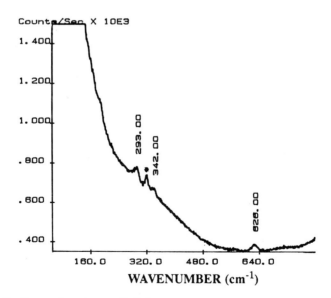

Figure 3: Raman Spectrum of CuO

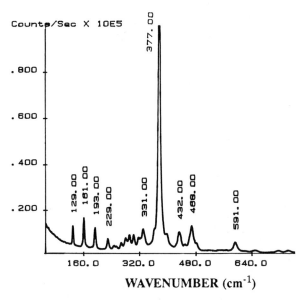

Figure 4: Raman Spectrum of Y_2O_3.

though some evidence exists for others. Spectra obtained from mixtures of 1:1 molar ratios of BaO_2, BaO, or $BaCO_3$ to Y_2O_3 and heated for more than eight hours at 950°C resembled that of Fig. 5(a). The strong 377 cm^{-1} peak of Y_2O_3 (Fig. 4) is used as a diagnostic peak indicating the extent of reaction completion. In Fig. 5(a) this peak is still evident, but it diminishes with successive heatings as the reaction becomes more complete.

Spectra obtained from 4:1 molar ratios of BaO, BaO_2, or $BaCO_3$ to Y_2O_3 (Fig. 5(b)) have no peaks in common with the spectra attributed to the 1:1 phase and obviously represent a different product. Further heating produced no obvious changes.

Mixtures of intermediate molar ratios (2 or 3 moles of barium compound to 1 Y_2O_3) resulted in Raman spectra indicative of mixtures of BaY_2O_4 and BaY_2O_7, with BaY_2O_4 predominating.

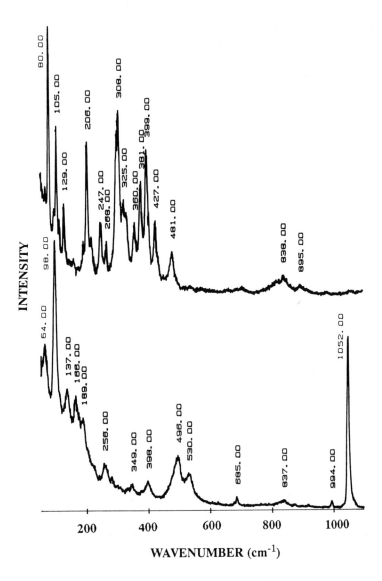

Figure 5: Raman Spectrum of (a) BaY_2O_4 and (b) $Ba_4Y_2O_7$.

3.2.2 Y_2O_3-CuO System

The blue-green phases which results from heating 2:1 molar ratios of Y_2O_3 with CuO is $Y_2Cu_2O_5$, the most stable phase in the Y_2O_3-CuO system (5-8). The Raman spectrum (Fig. 6) consists of several strong bands in the region from 170 to 640 cm^{-1} and a number of sharp peaks below 160 cm^{-1}. We have not observed $Y_2Cu_2O_5$ as a contaminating phase in 123 preparations.

3.2.3 BaO-CuO System

$BaCuO_2$ is well known (11-13); Ba_2CuO_3 and Ba_3CuO_4 were reported by Frase et. al. (5), but are hygroscopic and unstable in air. Six major peaks are visible in the Raman spectrum of $BaCuO_2$ (Fig. 7) at 215, 347, 440, 495, 583, and 635 cm^{-1}. Attempts to record Raman spectra of Ba_2CuO_3 or Ba_3CuO_4 were unsuccessful. In some of our spectra of this phase the peak at 635 cm^{-1} was more intense than the peak at 583 cm^{-1}.

3.3 Y_2O_3-BaO-CuO TERNARY SYSTEM

In addition to the most stable superconducting black 123 phase and the non-superconducting green 211 phase, at least three other metastable phases have been identified. The 132 phase (non-superconducting) has been prepared by several groups (5,14,15). McKittrick, et. al. have prepared $YBa_3Cu_4O_x$ and $Y_2Ba_5Cu_7O_x$ by means of rapid solidification. These two phases also exhibited superconductivity with T_c about 87.5K (16). We have obtained Raman spectra of the 123 and 211 phases.

As is now well-known (17), the crystal structure of $YBa_2Cu_3O_{6+x}$ depends on oxygen content. When x > 0.4 the material is orthorhombic and superconducting, but when x falls below 0.4 the crystal structure shifts to tetragonal and the material becomes semiconducting. The oxygen content, and hence the

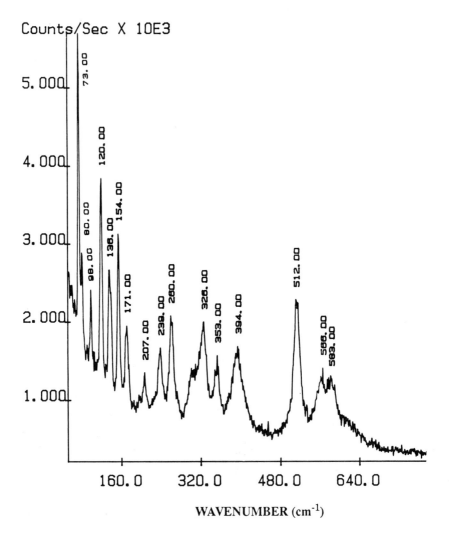

Figure 6: Raman Spectrum of $Y_2Cu_2O_5$

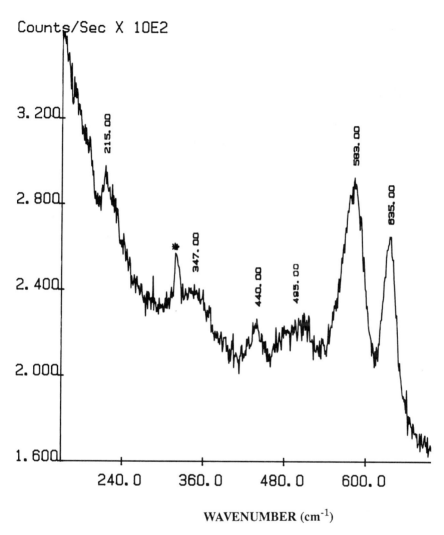

Figure 7: Raman Spectrum of $BaCuO_2$

superconductivity, is very sensitive to changes in the method of preparation and subsequent treatment. Both 123 cyrstal structures are black opaque materials with small Raman cross-sections, making their spectra relatively weak. Nevertheless, in the 100 to 800 cm^{-1} region six weak Raman bands at 145, 338, 434, 491, 582 and 636 cm^{-1} can be identified in the orthorhombic phase (Fig. 8(a)) and five at 140, 338, 468, 568, and 592 cm^{-1} in the tetragonal phase (Fig. 8(b)). The tetragonal phase was obtained by heating the orthorhombic sample in a vacuum at $400^{o}C$ for four hours. Annealing the tetragonal sample in oxygen restores superconductivity and original Raman peaks (Fig. 8(c)). The exact oxygen content for these samples was not determined, but their phases were confirmed by X-ray diffraction and resistivity measurements.

In contrast to the 123 phases, the 211 phase has a large Raman cross-section and at least seven strong lines can be identified in the 50 to 800 cm^{-1} region (Figure 9). These lines may interfere with or overwhelm the weaker lines of the 123 phase. This problem becomes particularly acute when a relatively large spot of a mixed phase pellet is sampled. For example, in the mixed spectrum for the 123 phase obtained by Rosen, et. al.(18), most 123 bands are obscured by the stronger 211 lines. Our 211 spectra are in very close agreement with those previously published (18,19).

Published spectra of the 123 phase (18-25), however, show a number of inconsistencies with 1 to 10 distinguishable peaks being reported. Rosen, et. al. (18) attributed all but two peaks in their published spectrum to contaminating phases and assisgned these (at 327 and 502 cm^{-1}) to the lattice vibrations of the orthorhombic 123 compound. Stavola, et. al (21) reported two peaks at 435 and 495 cm^{-1} for orthorhombic $YBa_2Cu_3O_{6+x}$. Hemley and Mao (19) reported four peaks at 142, 338, 483, and 595 cm^{-1} for the tetragonal phase. Using orthorhmobic 123 samples prepared in a manner similar to our own, Iqbal, et. al. (22) reported Raman bands at 430, 497, 583, and 632 cm^{-1}, which are in

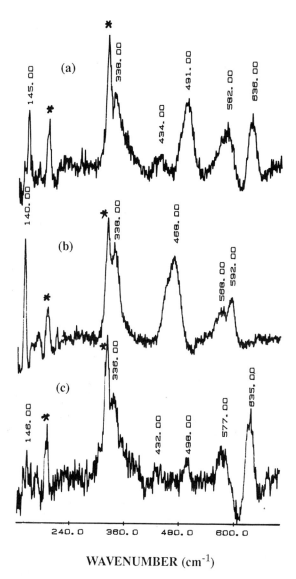

WAVENUMBER (cm⁻¹)

Figure 8: Raman Spectrum of (a) orthorhombic, (b) tetragonal, (c) orthorhombic "123" phase (background subtracted). Peaks marked with asterisks are artifacts of our optical system.

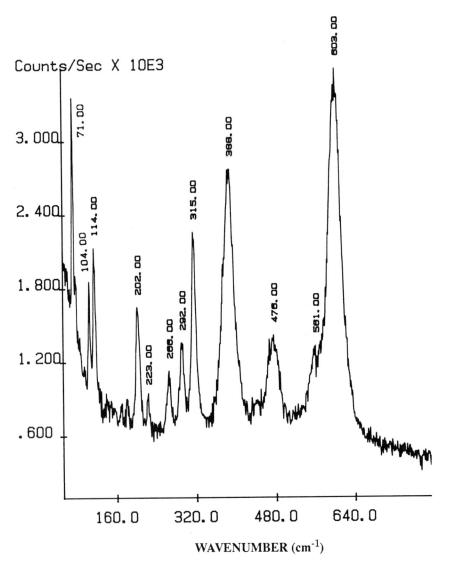

Figure 9: Raman Spectrum of the green "211" phase, Y_2BaCuO_5.

good agreement with our results with the exception of the 145 and 338 cm^{-1} peaks which did not appear in their published spectrum. McFarlane et. al. (24) reported Raman active phonon modes at 153, 217, 291, 309, 335, 441, 493, 506, 601, and 640 cm^{-1} for their orthorhombic samples. Spectra of the 300 to 900 cm^{-1} region reported by Dai, et. al. (25) show peaks at 344, 480, and 581 cm^{-1} for the tetragonal phase and peaks at 502, 589, and 636 cm^{-1} for the orthorhombic phase.

Spectra of other rare earth high T_c superconductors have also been reported. In their study of the Raman scattering of orthorhombic $YBa_2Cu_3O_{6+x}$, $EuBa_2Cu_3O_{6+x}$, and $SmBa_2Cu_3O_{6+x}$, Liu, et. al. (23) observed a profound difference in the band intensities in the region of 300 to 800 cm^{-1} of the spectra of three different 123 compounds. Only a weak band at 505 cm^{-1} is observed in the spectrum of the yttrium compound, while three weak bands at 432, 584 and 640 cm^{-1}, along with a strong 505 cm^{-1}, appeared in the Sm compound. In the spectrum of the Eu compound, two strong bands at 505 and 640 cm^{-1} and two medium bands at 432 and 584 cm^{-1} were observed. In their study of isotope effects, Batlogg, et. al. (20), reported peaks at 433, 501, 592 and 644 cm^{-1} for orthorhombic $EuBa_2Cu_3O_{6+x}$ at 77 K. Our Raman studies of Sm, Eu, and Nd orthorhombic "123" compounds show no substantial differences from the Raman spectra of orthorhombic $YBa_2Cu_3O_{6+x}$. These results are consistent with the similarity of the crystal structures of the various rare-earth copper oxides and with the assumption that the observed spectra arise from vibrations not involving the rare-earth ion.

The wide differences in the results are probably due to different methods of preparation and oxygen treatment of the superconductor phases or to the presence of other phases in the region sampled. From these studies and our own results, we conclude that there are six or seven major peaks in the Raman spectrum of orthorhombic $YBa_2Cu_3O_{6+x}$. In our work the peaks at about 145, 338, 434, 491, and 582 cm^{-1} appear consistently. The 636 cm^{-1} peak appears to depend on the sample preparation and treatment.

4. CONCLUSIONS

We have recorded Raman spectra of several phases that are most likely to occur as minor phases in the preparation of the superconductor $YBa_2Cu_3O_{6+x}$. These Raman spectra may be used to characterize micro-structures or -phases in the sintered materials. Contamination of high T_c superconductor materials by unreacted reagents, binary phases, or non-superconducting and low T_c superconducting ternary phases can lower the effective T_c of the superconducting material. Furthermore, if the critical current density of the high T_c superconducting oxide depends on the purity of phases, detection and characterization of undesirable phases will be essential for further development of high T_c superconductors. Raman microprobe spectroscopy can therefore provide a powerful tool in this development.

ACKNOWLEDGMENTS

Work at the University of Alabama in Huntsville was supported by NASA Grants NAG8-089 and NAG8-051, and by NSF Alabama EPSCoR Grant RII-8610669. The Authors thank Dr. D. Coble for helpful discussions. BHL thanks Universities Space Research Association (USRA) for a Visiting Scientist Fellowship at Marshall Space Flight Center.

REFERENCES

1. M.K. Wu, J.R. Ashburn, C.J. Torng, P.H. Hor, R.L. Meng, L. Gao, Z.J. Huang, Y.Q. Wang, and C.W. Chu, Phys. Rev. Lett. 58, 908 (1987); P.H. Hor, L. Gao, R.L. Meng, Z.J. Huang, Y.Q. Wang, K. Forster, J. Vassilious, C.W. Chu, M.K. Wu, J.R. Ashburn, and C.J. Torng, Phys. Rev. Lett. 58, 911 (1987).

2. R.M. Hazen, L.W. Finger, R.L. Angel, C.T. Prewitt, N.L. Ross, H.K. Mao, C.G. Hadidiacos, P.H. Hor, R.L. Meng, and C.W. Chu, Phys. Rev. B35 7238 (1987).

3. Y. LePage, W.R. McKinnon, J.M. Tarascon, L.H. Green, G.W. Hull, and D.W. Huang, Phys. Rev. B35 7245 (1987).

4. T. Siegrist, S. Sunshine, D.W. Murphy, R.J. Cava, and S.M. Zahurak, Phys. Rev. B35 7237 (1987).

5. K.G. Frase, E.G. Liniger, and D.R. Clarke, Comm. Am. Ceram. Soc. (to be published).

6. K.G. Frase and D.R. Clarke, Advanced Ceram. Mat. 2, 295 (1987)

7. R.S. Roth, K.L. Davis, and J.R. Dennis, Advanced Ceram. Mat. 2, 303 (1987)

8. G. Wang, S.J. Hwu, S.N. Song, J.B. Ketterson, L.D. Marks, K.R. Poeppelmeier, and T.O. Mason, Advanced Ceram. Mat. 2, 313 (1987).

9. W.P. Griffith, J. Chem Soc. (A) 286 (1970).

10. K.H. Rieder, B.A. Weinstein, M. Cardona, and H. Bilz, Phys. Rev. B8 4780 (1973).

11. M. Arjomand and D.J. Machin, J. Chem Soc. Dalton 1061 (1975).

12. H.N. Migeon, F. Jeannot, M. Zanne, and J. Aubry, Rev. Chim. Miner, 13, 440 (1976).

13. R. Kipka and H. Muller-Buschbaum Z. Naturforsch, 32B, 121 (1977).

14. R.S. Roth, J.R. Dennis, and K.L. Davis, to appear in Introduction to Phase Diagrams for Ceramists, Vol. 6 (in press).

15. G. Wang, S.J. Hwu, S.N. Song, J.B. Ketterson, L.D. Marks, K.R. Poeppelmeier, and T.O. Mason, J. Solid State Chem. Lett., (to be published).

16. J. McKittrick, L.Q. Chen, S. Sasayama, M.E. McHenry, G. Kalonji, and R.C. O'Handley, Adv. Ceram. Mat. 2, 353 (1987).

17. J.M. Tarascon, P. Barboux, B.G. Bagley, L.H. Green, W.

McKinnon, and G.W. Hull, in Chemistry of High-Temperature Superconductors, ed. D. Nelson, M.S. Whittingham, and T.F. George, American Chemical Society, P. 198 (1987).

18. H. Rosen, E.M. Engler, T.C. Strand, V.Y. Lee, and D. Bethune, Phys. Rev. B36, 726 (1987)

19. R.J. Hemley, and H.K. Mao, Phys. Rev. Lett. 58, 2340 (1987).

20. Batlogg, R.J. Cava, A. Jayaraman, R.B. van Dover, G.A. Kouroukis, S. Sunshine, D.W. Murphy, L.W. Rupp, H.S. Chen, A. White, K.T. Short, A.M. Mujsce, and E.A. Rietman, Phys. Rev. Lett. 58, 2333 (1987)

21. M. Stavola, D.M. Krol, W. Weber, S.A. Sunshine, A. Jayaraman, G.A. Kourouklis, R.J. Cava, and E.A. Rietman, Phys. Rev. B36, 850 (1987).

22. Z. Iqbal, S.W. Steinhauser, A. Bose, N. Cipollini, and H. Eckhardt, Phys. Rev. B36, 2283 (1987).

23. R. Liu, R. Merlin, M. Cardona, Hj. Mattausch, W. Bauhofer, A. Simon, F. Garcia-Alvarado, E. Moran, M. Vallet, J.M. Gonzalez-Calbet, and M.A. Alario, Solid State Comm. 63, 839 (1987).

24. R.M. Macfarlane, H.Rosen, and H. Seki, Solid State Comm. 63, 831 (1987).

25. Y. Dai, J.S. Swinnea, H. Steinfink, J.B. Goodenough, and A. Campion, J. Am. Chem. Soc. 109, 5291 (1987).

Index

A

A15 compounds, 88, 228

Absorption edge, free
carrier, 286

Acoustic modes, 223

Ag_2F, 116

Ag_7O_8X:
formulation of, 10
oxidation state
of Ag, 10
structure of, 10
T_c, 10

Alkali metals, 111

Alkaline earth, 110

Alloys, superconducting, 38

Anharmonic motion, of Cu^{+2}
ion, 112

Anion stoichiometry,
in $YBa_2Cu_3O_{7-\delta}$, 141

Anion vacancies, 215

Antibonding character,
of wave functions,
113, 116

Antibonding state, 109

Antiferromagnetism,
in La_2CuO_4, 176, 211

Antiferromagnetic spin
fluctuations, 81, 85

Antiferromagnetic state,
226

Applications of supercon-
ductors, 74, 251

Arrhenius plot, 245

Atom-atom potential, 289
calculations on La_2CuO_4,
290

Atomic coordinates:
for $La_{1.85}Sr_{0.15}CuO_4$,
220

A_xMoO_3, 4

A_xReO_3, 5

A_xWO, 4

$A_xWO_{3-x}F_x$, 148

B

$Ba(Pb,Bi)O_3$, 25

   structure of, 6

   T_c, 6

$Ba_2Bi^{III}Bi^VO_6$, 26, 229

   pervoskite structure
     of, 7

   real charges in, 24

$Ba_2Cu_3O_7$ layer, 191

$Ba_2Cu_3O_{7-y}{}^{3-}$ slab, 181,
    184, 194

   and structure of, 190

$Ba_2Cu_3O_7{}^{3-}$ slab, 195, 196

Ba_2CuO_3, 350, 357

Ba_2MgWO_6, structure of, 26

$Ba_2Y_2O_5$, 349

$Ba_2YCu_3O_{6.2}F_{0.63}$, resistivity data for, 146

Ba_3CuO_4, 350, 357

$Ba_3Y_4O_9$, 349

$Ba_4Y_2O_7$, 351

   Raman spectrum of, 356

$Ba_4Y_2O_7$, 349

$BaBi_{1-x}Pb_xO_3$, 229, 230

$BaBiO_3$, 25, 26

   2p levels of oxygen
     in, 25

   charge density wave
     in, 26

   decomposition tempera-
     ture, 6

   disproportionation in,
     25

   formulation of, 6

   monoclinic symmetry 8

   semiconducting proper-
     ties of, 6

$BaCO_3$, Raman spectrum
   of, 352

$BaCuO_2$, 338, 350

   in BaO-CuO system, 357

   preparation of, 350

   Raman spectrum of, 359

Band, of CuO_3 chain, 194

Band structure,

   of $Ba_2Cu_3O_{7-y}{}^{3-}$ slab,
    190

   of CuO_2 layer, 189

   and electronic transi-
    tions, 275

   of $Ga_2Cu_3O_{7-y}{}^{3-}$ slab,
    199

   of high T_c supercon-
    ductors, 108

   of La_2CuO_4, 13

Band structure calcula-
   tions, and O1 atom
    displacement, 200

Band, x^2-y^2, 189, 190, 194

BaO, Raman spectrum of, 353

$BaO-Y_2O_3$ system, 355

$BaPb_{0.75}Bi_{0.25}O_3$, T_c, 9

$BaPb_{1-x}Bi_xO_3$:

   disproportionated state,
    26

   isotope effect in, 82

   phase diagram for, 7

   positron trapping in,
    305

   properties, 6

s^1 cation in, 9

semiconducting, 10

structure of, 8

T_c, 8

$BaPbO_3$, 26

density of states in, 7

metallic properties, 6

orthorhombic symmetry

of, 8

T_c, 6

$BaPh_{0.7}Bi_{0.3}O_3$, 58

BaY_2O_4, 349, 351

Raman spectrum of, 356

BCS energy gap, 43, 89

BCS gap parameter, 245

BCS pairing, 81

BCS theory, 51, 83

and disproportionation

mechanism, 32

and electronic structure

gap, 43

and the isotope effect,

223

and $LiTi_2O_4$, 5

and oxide superconduc-

tors, 23

and tungsten bronzes, 4

BCS vacuum polarization,

101

Binding energy, 102

Bipolarons, 113, 230

Bogoliubov transformation,

101

Bond distances:

Cu-O, 12, 15

Cu-O in $YBa_2Cu_3O_{7-y}$, 185

Bose condensation, 80

Bose-Einstein condensation,

88

Broad band electronic

applications, 159

C

$Ca_2MnO_{4-x}F_x$, 147

Calcination conditions, 127

Calcination method, for

preparing $YBa_2Cu_3O_{6+x}$,

349, 350

Calcined precipitates, 127

Calcining, of oxalate

samples, 155

Cation contamination, 124

Cation-anion mixing, 113

Ceramics, oxygen-depleted

YBCO, 243

Charge density wave, 26,

229, 230

in $BaBiO_3$, 26

Charge distribution, 168

Charge transfer excita-

tions, 80 ff., 113

Chemical titration, 168

vs. XANES, 173

Chlorides, 117

Coherence length, 228, 239

temperature-dependent,

48

Condensation energy, 46

Conductance:

   of $YBa_2Cu_3O_{6.4}$, 246, 247

Conduction band, 3

   in NbO, 3

   π^*, 4

   Pb 6s, 6

   σ^*, 4

   in $SrTiO_{3-x}$, 4

   in TiO, 3

Conduction electrons,

   mean free path for, 111

Conductivity:

   frequency dependent,
     277

   of $La_{1.83}Sr_{0.17}CuO_4$,
     215

   of $La_{1.85}Sr_{0.15}CuO_4$,
     278

   optical, 89

Contact resistance, 161,
   251, 252

Contacts, 252

   metal-superconductor,
     161

   semiconductor-super-
     conductor, 161

Cooper pairs, 50, 101, 102

   and coherence length, 86

   as current carriers, 227

   formation of, 204

   and isotope effect, 223

   and Josephson effect, 54

   and s-wave symmetry, 80

   and superconductivity
     mechanism, 23

and tunneling, 236

Coordination, of copper,
   184, 328

Copper:

   3d-block levels in, 187

   contacts, 254

   oxidation states, 184

   oxides, 107

   valence, 167

   valence, in La_2CuO_4, 171

Coprecipitation,

   of carbonates, 214

Coulomb correlations, 109

Covalency, 168

   of Cu-O bonds, 29

Covalent bonding, 18

Creation operators, 101

Critical current, 239

Critical current density,
   86, 159, 161

   of $YBa_2Cu_3O_{6+x}$ film, 162

Critical field, 86, 228

   temperature dependence
     of, 46

   of $YBa_2(Cu_{0.95}Ni_{0.05})_3$-
     $O_{7-\delta}$, 137, 138

Critical point, 102

Cryostat, 261

Crystal structures:

   evolution with tempera-
     ture, 212

   simulation of, 289

$Cs_2[Au^ICl_2][Au^{III}Cl_4]$, 229

$Cs_4[Sb^{III}Cl_6][Sb^VCl_6]$, 229

Cu-O bonds:

covalency of, 29

in $LaCuO_4$, 29

and T_c, 30

Cu-O chains

in high T_c superconductors, 107, 112, 167, 305

random orientation of, 20

Cu-O planes, 167

Cu-O sheets, in high T_c superconductors, 107, 112

$[Cu-O]^+$ complexes:

concentration of, 90

and mobile holes, 88

Cu-O-Cu bonds, 12

CuO, 25

2p levels of oxygen in, 25

Raman spectrum of, 354

CuO_2 chain, 193

CuO_2 layer, 229

in $BaCu_3O_7^{3-}$ slab, 181

band structure of, 189

Cu 3d orbital contributions, 196

density of states for, 193

dispersion relation of x^2-y^2 band, 190

O 2p orbital contributions, 196

x^2-y^2 band in, 194

orthorhombic distortion

in, 228

in $YBa_2Cu_3O_{7-y}$, 203

CuO_2 sheets, 14

coordination of copper in, 16

Cu-O bonding in, 15

Cu^{II} in, 17

in La_2CuO_4, 14, 15

in 1·2·3 compounds, 15

in R_2CuO_4, 28

CuO_3 chain:

absence in tetragonal phase of $YBa_2Cu_3O_{7-y}$, 183

in a $Ba_2Cu_3O_{7-y}^{3-}$ slab, 190

coordination of copper in, 181

x^2-y^2 band in, 192

z^2-y^2 band in, 194

CuO_4 layer:

x^2-y^2 band in, 189

in $La_{2-x}Cu_xO_4$, 194

in $La_{2-x}M_xO_4$, 203

Current-voltage characteristics, of $Y_1Ba_2Cu_3O_{7-x}$, 237

D

d-block bands, 187

dispersion relations, 192, 196

Debye energies, 85

Debye frequency, 52, 53, 84

Debye temperature, 53

Demagnetization factor, 55

Density functional theory, 99

Density of states, 4, 23, 52

  of d-block bands, 193, 197

  at Fermi level, 4 ff.

  at Fermi surface, 80, 82

  in $La_{1.85}Sr_{0.15}CuO_4$, 225, 227

  phonon, 224, 225

  of x^2-y^2 band, 191

Diamagnetic shielding

  of $GdBa_2Cu_3O_{7-x}$, 314

Diamagnetism, 45

  fractional ideal, 281

Differential thermal

  analysis of

  $YBa_2Cu_3O_{7-\sigma}$, 337

Diffuse magnon modes, 113

Dilatometer densification

  data for $YBa_2Cu_3O_{7-\sigma}$, 339

Disporportionation mecha-

  nism for superconduc-

  tivity, 28-31

Disproportionation: 175

  in copper oxides, 27

  mechanism for super-

  conductivity, 28-31

Disproportionation mecha-

  nism:

  d^1 cations, 31

  d^8 cations, 27

d^9 cations, 27

d^{10} cations, 27

Doped phases,

  $YBa_2(Cu_{1-x}M_x)_3O_{7-\delta}$, 133

Doping:

  and copper valence, 172

  of La_2CuO_4, with Sr, 172

  with magnetic elements, 132

  with nonmagnetic

  elements, 132

  and oxygen oxidation, 173

  in $R_{2-x}Sr_xCuO_4$, 29, 216

  in $SrTiO_3$, 3

Drude term, 286

E

Effective mass,

  of carriers, 88

Electrical resistance:

  dc, 268

  determination of, 328

Electrical resistivity,

  determination of, 350

Electron localization, 13

Electron microscopy,

  of $YBa_2Cu_3O_{7-z}$, 267

Electron pairing, 23, 26, 240

Electron-electron attrac-

  tion, 52

Electron-electron interac-

  tion, 13

Electron-hole pairs, 81, 84

Electron-phonon coupling,
58, 99, 222

Electron-phonon coupling
constant, 83

Electron-phonon interac-
tion, 32, 80, 104, 113,

Electronegativity differ-
ence,
between anion and
cation, 114

Electronic excitations, 280

Electronic polarization,
84, 90

Electropositive cations,
influence of, 110

Ellipsometry, 276

Energy gap, 103, 286
BCS, 43, 89
spectroscopic, 283

EPR:
of $GdBa_2Cu_3O_{7-x}$, 319
of $YBa_2Cu_3O_{7-z}$, 267
of $YBa_2Cu_3O_{7-z}$, 271

$EuBa_2Cu_3O_{6+x}$,
Raman scattering of, 363

Exciton, 54, 80, 81, 89,
113

Exciton mechanism, 85

Exciton theories, 113

Extended Huckel calcula-
tions, 190

F

Fabrication methods, 72

Fast ion conductor, 111,
115

Fermi contact interaction,
103

Fermi level, 104
density of states at, 4,
5, 23
and π^* band, 12
and z^2-y^2 band, 199

Fermi liquid, 86, 100

Fermi sea, 85, 100

Fermi surface, 52, 196
of d-block bands, 198
of z^2-y^2 band, 199

Ferroelastic materials, 19

Ferroelectric materials, 19

FIR reflectivity, 89

Fluorides, 116

Fluorination, of
$YBa_2Cu_3O_{7-\delta}$, 142

Flux exclusion, 44

Flux quantization, 236

Flux vortices, 88
pinning of, 88

Fluxons, 86

Formal valence states, 110,
114, 116, 167

Four-point resistance, 254

Four-terminal resistance,
262

Free electron model, 100

Frustration, in La_2CuO_4, 32

G

G-L parameter, 86

$Ga_2Cu_3O_{7-y}{}^{3-}$ slab,
  band structure of, 199
Gap theories, 88
$Gd_3Ba_3Cu_4O_z$, 268
$GdBa_2Cu_3O_7$, infrared
  reflectance of, 283
$GdBa_2Cu_3O_{7-x}$:
  EPR of, 319
  magnetic anomalies in,
    313
  magnetic susceptibility
    of, 313
$GdBa_2Cu_3O_{7-z}$:
  resistance of, 331
  X-ray powder diffrac-
    togram for, 329
$GdBa_2Cu_3O_x$, 328
  preparation of, 328
  T_c, 330
Gibbs chemical potential,
  103
Ginzburg-Landau parameter,
  49, 86
Glassy phases, 324
Gold contacts, 257
Grain boundaries, 9
Graphite, intercalated, 116

H
Hall effect studies, on
  $GdBa_2Cu_3O_{7-z}$, 330
Heat capacity, determi-
  nation of, 323
Heavy Fermion supercon-
  ductors, 82

Heine-Pippard mechanism,
  104
High frequency applica-
  tions, 159
$HoBa_2Cu_3O_{6+x}$, 69
  T_c, 69
Holes, 88, 90
Hopping energy, 246
Hopping processes, 245
Hydroxides, 123

I
I-V characteristic, 237
Incipient superconductor,
  243
Incoherent neutron scattering, 211, 223
Inductive effects, on
  Cu^{II}-O covalency, 28
Inelastic neutron scattering, 211, 223
Infrared frequencies:
  for $La_{2-x}M_xCuO_4$, 295
  for La_2CuO_4, 295
Infrared reflectance:
  of $GdBa_2Cu_3O_7$, 283
  of $RBa_2Cu_3O_7$, 285
  of $TmBa_2Cu_3O_7$, 283
  of $YBa_2Cu_3O_7$, 283
Infrared spectra, of
  $La_{2-x}M_xCuO_4$, 289
Infrared vibrations, in
  La_2CuO_4, 294
Intercalation reaction, 18

Interfacial superconduc-
tivity, 60
Intergrowth phases, 13
Interlayer interaction, 194
Intermediate state, 55
Intermetallic compounds, 1,
111
Iodometric titrations, 167
Ionic diffusion, 115
IR active vibrations:
frequencies of, 294
symmetries of, 294
IR inactive modes, 299
Isotope effect, 23, 51, 85
in $BaPb_{1-x}Bi_xO_3$, 82
in $La_{2-x}Sr_xCuO_4$, 82
in $YBa_2Cu_3O_{7-\delta}$, 82

J
Jahn-Teller effect, 58,
105, 112
Jahn-Teller ion, 327
Josephson effect, 54, 235
Josephson junction, 55, 61,
73, 238
Josephson supercurrent, 238
Josephson tunnel junction,
79

K
$K_{0.3}ReO_3$:
structure of, 5
T_c, 5
K_2NbO_3F, 147
K_2NiF_4, 10, 202, 216

Knight shift, 104
Kramers-Kronig analysis,
89, 275
K_xMoO_3, structure of, 5

L
$La_{1.25}Ba_{1.75}Cu_3O_{7.1}$, 20
structure of, 20
T_c, 20
$La_{1.83}Sr_{0.17}CuO_4$:
conductivity of, 215
$La_{1.85}Ba_{0.15}CuO_4$, 67
T_c, 60
$La_{1.85}Ca_{0.15}O_4$, 60
$La_{1.85}M_{0.15}CuO_4$, 228
$La_{1.85}Sr_{0.15}CuO_4$:
atomic coordinates, 220
conductivity of, 278
crystal structure, 221
current-voltage ex-
periments, 236
density of states in,
225, 227
electron-phonon interac
tion in, 223
lattice vibrations in,
211
neutron diffraction
profiles, 217
neutron diffraction
structure, 220
orthorhombic distortion
and superconduc-
tivity, 222
phonon density of states

in, 224, 225

reflectance of, 277

structure of, 211

titrations of, 169

unit cell parameters
for, 219

XANES of, 173

$La_{1.8}Ba_{0.2}CuO_{3.9}F_{0.11}$, 145

$La_{1.8}Ba_{0.2}CuO_{3.9}$, 145

resistivity of, 146

$La_{1.8}Ba_{0.2}CuO_4$, 221

$La_{1+x}Ba_{2-x}Cu_3O_7$, 109

$La_{1+x}Ba_{2-x}Cu_3O_y$:

aluminum substitution
in, 22

conductivity of, 22

fluoride substitution
in, 22

iron substitution in, 22

structure of, 22

$La_2-_xB_{ax}CuO_3F$, 142

$La_{2-x}A_xCuO_4$, 10, 13

structural transition
in, 14

$La_{2-x}A_xCuO_{4-z}$, 267

$La_{2-x}Ba_xCuO_4$, 58

$La_{2-x}Ba_xCuO_{4-z}$, 328

$La_{2-x}CoO_4$, 170

$La_{2-x}CuO_{4-x}$, 12

$La_{2-x}M_xCuO_4$, 107, 194, 213

infrared frequencies,
295

infrared spectra of, 289

preparation of, 213

Raman frequencies

for, 292

Raman spectra of, 289

$La_{2-x}M_x^{II}CuO_4$, 229

$La_{2-x}M_x^{II}CuO_{4-y}$, 212, 228

$La_{2-x}M_xO_4$, CuO_4 layers in,
203

$La_{2-x}Sr_xCu_3O_{7-\delta}$, 88

$La_{2-x}Sr_xCuO_4$, 167, 229

isotope effect in, 82

Meissner effect in, 281

optical properties of,
275

reflectance of, 279

XANES data for, 174

$La_{2-x}Sr_xCuO_{4-x/2}$, 171

$La_{2-x}Sr_xCuO_{4-y}$, 215

La_2CuO_4:

antiferromagnetism
in, 13

atom-atom potential
calculations, 290

band structure, 13

Cu^{2+} and Cu^{3+} in, 172

CuO_2 sheets in, 15

doping of, 222

frustration in, 32

infrared frequencies,
295

infrared vibrations, 294

lanthanum deficiency
in, 12

lattice vibrations in,
211

magnetic behavior of,
212

as a Mott insulator, 13

oxidation state of Cu
   in, 169

oxygen deficiency in, 12

peroxide formation in,
   170

phonon density of states
   in, 224, 225

Raman frequencies for,
   292

Raman vibrations in, 291

semiconducting proper-
   ties of, 13

spin density wave in, 13

structure of, 10, 11,
   14, 215

superconductivity in,
   13, 171

titrations of, 169

XANES of, 171, 173

La_2CuO_{4+x}, 170

La_2CuO_{4-x}, 12

$La_{3-x}Ba_{3+x}Cu_6O_{14}$, 107

$La_4BaCu_5O_{13+x}$, 213

$(La,A)_2CuO_4$, 6

Lagrange multiplier, 100

Lanthanide oxyfluorides,
   142

$(La,R,Sr)_2CuO_4$, 30

   electronegativity of R,
      30

   ionicity of Cu-O bands,
      30

   structure of, 30

   T_c, 30

Lattice parameters,
   for $YBa_2Cu_3O_{7-\delta}$, 133

Lattice polarizability, 115

Lattice vibrational modes,
   200

Lattice vibrations,
   in $La_{1.85}Sr_{0.15}CuO_4$, 211
   in La_2CuO_4, 211

Le Chatlier's principle,
   100

$Li_{1+x}Ti_{2-x}O_4$, 5

$Li_{1+x}Ti_{2-x}O_4$, 58

Lifetimes, of positrons,
   307

Linkages, 184, 194, 203

Liquid phases, 339, 341

$LiTi_2O_4$, 109
   structure of,
   T_c, 5

$LnBa_2Cu_3O_{7-x}$, 131

$LnBa_2Cu_3O_{7-x}F_x$, 142

Localized singlet pairs,
   113

London penetration depth,
   47

London theory, 46

Lorentzian oscillators, 286

Low-dimensional aspect, 112

M

Magnetic anomalies:
   in $GdBa_2Cu_3O_{7-x}$, 313
   in $YbBa_2Cu_3O_{7-x}$, 313

Magnetic excitations, 226

Magnetic field, critical,
    45, 46
Magnetic shielding appli-
    cation, 159
Magnetic susceptibility,
    and electronic density
        of states, 132
    of $GdBa_2Cu_3O_{7-x}$, 313
    of $YbBa_2Cu_3O_{7-x}$, 313
Magnetism and supercon-
    ductivity, competition
    between, 212
Magnetization, 44
Magnetization, of
    $YbBa_2Cu_3O_{7-x}$, 319
Magnon modes, 113
Magnons, 226
$MBa_2Cu_3O_7$, 107
$MBa_2Cu_3O_{7-y}$, 228
Mechanism:
    disproportionation, 28,
        31, 32
    of high temperature
        superconductivity, 80
    of pairing, 105
    for superconductivity,
        electron-lattice
        interactions, 6, 23,
        28
Meissner effect, 44, 50,
    102, 132, 275
    in $La_{2-x}Sr_xCuO_4$, 281
    in $YBa_2(Cu_{1-x}Ni_x)_3O_{7-\delta}$,
        137
    relationship to specific

heat, 321
Metal oxalate compounds,
    154
Metal-insulator transition,
    115, 117
Metal-metal distance:
    in NbO, 3
    in TiO, 3
Metal-metal interaction, 3
Metastable materials,
    $BaPb_{1-x}Bi_xO_3$, 8
Microelectronic
    applications, 244
Microelectronics, 251
Microstructure:
    of superconducting
        ceramics, 122
    of $YBa_2Cu_3O_{7-\delta}$, 335
    of $YBa_2Cu_{3\cdot09}O_{7-\sigma}$, 342
    $M^{III}Ba_2Cu_3O_{7-\delta}$, 212
    $M^{III}Ba_2Cu_3O_{7-\delta}$, 228
Mixed state, 55, 56
Mixed valent behavior, 167
Mixed-oxide synthesis, 121
Mixed-valence
    $Cu(II)-Cu(III)$, 327
Mixed-valence metals, 58
Mixing:
    between z^2-y^2 and x^2-y^2
        bands, 195
    of Cu3d and O2p bonds,
        148
Molybdenum bronzes:
    cubic, 5
    metallic, 4

Mott insulator, 13
Mott transition, 108, 111

N
$NaCuO_2$,
    formal charges in, 111
Nb_3Ge, T_c, 38
NbO, 1
    metal-metal distance
        in, 3
$Nd_{1.9}Sr_{0.1}CuO_4$, 15
    resistivity of, 15
Nd_2CuO_4, structure of, 15
Neutron diffraction:
    of $La_{1.85}Sr_{0.15}CuO_4$, 211
    of Nd_2CuO_4, 15
    showing oxygen thermal
        amplitudes, 111
    of Pr_2CuO_4, 15
    showing disorder in
        $YBa_2Cu_3O_x$, 21
Neutron diffraction
    profiles,
    of $La_{1.85}Sr_{0.15}CuO_4$, 217
Neutron diffraction
    structure,
    of $YBa_2Cu_3O_{7-y}$, 185
Neutron diffraction
    studies,
    on $La_{1.85}Sr_{0.15}CuO_4$, 220
Neutron scattering,
    incoherent inelastic,
        211, 223
Nickel compounds, 114
Nitrides, 117

Nuclear spin relaxation
    rate, 89

O
Onnes, H. Kammerlingh, 37
Optical properties, of
    $La_{2-x}Sr_xCuO_4$, 275
Optical transmission
    measurements, 89
Orbital interactions,
    between copper and
        oxygen, 186
Order parameter, phase of,
    235
Orthorhombic distortion,
    in $La_{1.85}Sr_{0.15}CuO_4$,
        211, 216, 220, 221
Orthorhombic-to-tetragonal
    phase transition, 182
Oscillator strength:
    of electronic modes, 281
    of vibrational modes,
        281
Oxalate precipitation
    methods, 153
Oxalic acid, use in
    synthesis, 154, 156
Oxidation number, 328
Oxidation:
    of copper, 110
    of oxygen, 110
Oxidation state:
    of Ag in Ag_7O_8X, 10
    of Bi, 9
    of Cu, 167, 174, 186

of Sb, 9
of Sn, 9
Oxides, 114
   metallic, 4
   nonsuperconducting, 109
   ternary, 3
Oxyfluorides, 141
   Ln-Ba-Cu, 142
   synthetic methods, 145
Oxygen:
   atom vacancies, 183, 202
   2p-block levels in, 187
   atoms, quantum
     tunnelling of, 325
   content, dependence of
     T_c on, 200, 201
   content, of CuI atom
     plane, 202
   defects, in
     $La_{2-x}Sr_xCuO_4$, 171
   deficiency, in
     $RBa2Cu_3O_x$, 19
   deficiency, in La_2CuO_4,
     12
   stoichiometry, 19
   stoichiometry, in
     $YBa_2(Cu_{1-x}Ni_x)_3O_{7-\delta}$,
     136
   vacancies, 19
   vacancies, ordering of,
     336, 343
   vacancies, positron
     trapping at, 305, 307
Oxygen-depleted YBCO, 243

P
p-block levels, 187
Pairing mechanisms, 81
Pairing, of carriers, 79
Paramagnetism, 45
Pauli principle, 80
PbO_2, 109
Peierls distortion, 289
Peierls transition, 11
Penetration depth, 44
Penetration length, 86
Perfect diamagnetism, 44,
   46
Perovskite, 68, 107, 110,
   131
   oxygen deficient, 212
   structure, 3
Perovskites:
   A_xMoO_3, 2, 4, 5
   A_xReO_3, 2,5
   A_xWO_3, 2, 4
   $Ba(Pb,Bi)O_3$, 6, 7, 24
   $SrTiO_{3-x}$, 2, 3
Peroxide formation, in
   La_2CuO_4, 170
Persistent current, 41
Phase diagram,
   for R_2CuO_4, 29
   for ternary oxide
     system, 351
   for Y_2O_3-BaO-CuO system,
     354
Phase equilibria, 349
Phase transition,
   in La_2CuO_4, 12

of Fermi liquid, 101
second order, 104
Phonon density of states,
    224, 225, 227
Phonon dispersion, 223
Phonon mechanisms, 113
Phonon modes, 284
Phonon-mediated pairing,
    80, 84
    evidence against, 80
Phonons, 81
Pippard coherence length,
    48, 86
Pippard model, 47
Plasma frequency, 286
Plasmon-phonon modes, 284
Plasmons, 80, 113
Polarization, of electrons,
    84, 90
Porosity, of
    $YBa_2Cu_3O_{7-\delta}$, 337, 343
Positron annihilation
    experiments, 85
Positron annihilation
    lifetime spectroscopy,
    305
Positron lifetime, 85
Positron trapping, 305
$Pr_{1.9}Sr_{0.1}CuO_4$, 15
    resistivity of, 15
Pr_2CuO_4, 15
    structure of, 15
Printing pastes, 161
Processing, 159
    of $GdBa_2Cu_3O_{7-z}$, 327

of $YBa_2Cu_3O_{7-\sigma}$, 335
Processing routes,
    influence on
    microstructure, 122
$Pt^{II}(C_2H_5NH_2)_4][Pt^{IV}-$
    $(C_2H_5NH_2)_4Cl_2]Cl_4$, 229

Q
Quantum tunnelling, of
    oxygen atoms, 325
Quaternay copper oxides,
    211

R
$R_{2-x}Sr_xCuO_4$, 29
R_2CuO_4, 29
    R = Pr, Nd, Sm, Gd, 14
    structure of, 11, 14
Raman active modes:
    frequencies of, 291
    symmetries of, 291
Raman frequencies:
    of $La_{2-x}M_xCuO_4$, 292
    of La_2CuO_4, 292
Raman inactive modes, 299
Raman measurements, 351
Raman microprobe
    spectroscopy, 364
Raman scattering:
    of $EuBa_2Cu_3O_{6+x}$, 363
    of $SmBa_2Cu_3O6_{+x}$, 363
    of $YBa_2Cu_3O_{6+x}$, 363
Raman spectra:
    of binary systems, 354
    of $La_{2-x}M_xCuO_4$, 289

and secondary phases,
  350
of ternary systems, 357
Raman spectrum:
  of $Ba_4Y_2O_7$, 356
  of $BaCO_3$, 352, 353
  of $BaCuO_2$, 359
  of BaO, 353
  of BaY_2O_4, 356
  of CuO, 354
  of orthorhombic 123
    phase, 361
  of tetragonal 123 phase,
    361
  of Y_2BaCuO_5, 362
  of $Y_2Cu_2O_5$, 358
  of Y_2O_3, 355
Rare earth cations,
  magnetic, 131
$RBa_2Cu_3O_6$, 17
  CuO_2 sheets in, 16, 17
  semiconducting, 17
$RBa_2Cu_3O_7$, 6
  infrared reflectance
    of, 285
$RBa_2Cu_3O_{7-x}$, defects
  in, 32
$RBa_2Cu_3O_{7-z}$, 267
$RBa_2Cu_3O_x$, 15
  oxygen deficiency in,
    19
$RBa_2Cu_3O_x$:
  structural changes in,
    19
  structure of, 15

Reflectance:
  concentration dependence
    of, 278
  infrared, 283
  of $La_{1.85}Sr_{0.15}CuO_4$, 277
  of $La_{2-x}Sr_xCuO_4$, 279
Reflectivity spectra,
  of $La_{2-x}Sr_xCuO_4$, 89
Reflectivity studies, 90
ReO_3, structure of, 5
Resistance:
  determination of, 328
  of $GdBa_2Cu_3O_{7-z}$, 331
  of $YBa_2Cu_3O_{6.4}$, 244
  of $YBa_2Cu_3O_{6.4}$, 245
  of $YBa_2Cu_3O_{6.9}$, 244
  of $YBa_2Cu_3O_{6+x}$ film, 163
  of $YBa_2Cu_3O_{7-z}$, 271
Resistivity:
  of $La_{1.8}Ba_{0.2}CuO_{3.9}$, 146
  electrical, 37, 39
  electrical,
    loss of, 38, 40
  of $Ba_2YCu_3O_{6.2}F_{0.63}$, 146
  of $YBa_2(Cu_{1-x}Ni_x)_3O_{7-\delta}$,
    134, 135
  temperature dependence
    of, 111
Resonant valence bond
  theory, 88
Resonating valence bond, 81
Rhenium bronzes, 5
Rietveldt method, 218
Rigid band approximation,
  192

S

s-Wave symmetry, 80

s^1 cation, 9

Scanning electron
   micrographs, of 1,2,3
   samples, 128

Screen printing techniques,
   161

Self consistent field
   calculations, 204

SEM, 269

Shapiro steps, 79, 238, 240

Silk-screen method, 160

Silver oxides, 10

Silver paint, 268

Singlet pairs, 113

Sintering studies, 127
   on $YBa_2Cu_3O_{7-\sigma}$, 336

$SmBa_2Cu_3O6_{+x}$, Raman
   scattering of, 363

Soft modes, 226

Soft plasmons, 80, 81, 85

Sol-gel techniques, 214

Solid state fusion, 69

Solution method,
   for synthesis, 72

Sommerfeld constant, 82

Specific heat, 101, 102
   of $(La_{0.91}^-$
      $Sr_{0.09})_2CuO_{4-y}$, 322
   of $(La_{0.9}Ba_{0.1})_2CuO_{4-y}$,
      322
   of $La_{1.85}Ba_{0.15}CuO_{4-y}$,
      322

of $La_{1.85}Sr_{0.15}CuO_{4-y}$,
   322

of $La_{1.8}Ba_{0.2}CuO_{4-y}$,
   322

of $La_{1.8}Sr_{0.2}CuO_{4-y}$,
   322

and normal electrons,
   321

temperature dependence
   of, 41, 42, 88

of $YBa_2Cu_3O_{6+\delta}$, 324

of $YBaCu_3O_{6+\delta}$, 321

Spectroellipsometry, 276

Spectroreflectometry, 276

Spectroscopic energy gap,
   283

Spectroscopy, Raman
   microprobe, 354

Spin density wave, 13, 226

Spin fluctuations, 80

Spin susceptibility, 102

Spin-bipolaron formation,
   81

Spinel structure, 5

SQUIDS, 61

$Sr_2FeO_{3+x}F_{1-x}$, 147

Sr_2FeO_3F, 148

$SrTiO_3$, stoichiometric, 3

$SrTiO_{3-x}$:
   covalent bonding in, 3
   metal-metal distance
      in, 3

Strongly coupled super-
   conductors, 52

Structures:

of 1:2:3 compounds, 18

of $La_{1.85}Sr_{0.15}CuO_4$, 211

of La_2CuO_4, 11

of $La_4BaCu_5O_{13}$, 109

of NbO, 3

of R_2CuO_4, 11

of TiO, 3

of $YBa_2Cu_3O_6$, 16

of $YBa_2Cu_3O_{6+x}$, 357

of $YBa_2Cu_3O_7$, 16

of $YBa_2Cu_3O_{7-x}$, 131

of $YBa_2Cu_3O_{7-y}$, 181

of $YBa_2Cu_3O_{7-y}$, 182

Sulfides, 117

Supercomputers, 73

Superconducting film, 72, 73

Superconducting gap, 50

Superconducting materials, processing of, 60

Superconducting mechanism:

bipolaron, 80, 81, 113

characteristics of, 113

charge transfer excitation, 80, 81

and disproportionation, 175

exciton, 80, 81, 113

excitonic enhanced, 85

magnetic, 80, 81, 113

phonon, 80, 81, 113

plasmon, 80, 81, 113

Superconducting properties:

of Ba(Pb, Bi)O_3, 8, 9, 10

of La_2CuO_4, 13

of $R_{2-x}Sr_xCuO_4$, 29

of $RBa_2Cu_3O_6$, 17, 19, 22

of $YBa_2Cu_3O_{7-\delta}$, 87

of $YBa_2Cu_3O_xF_y$, 143

Superconducting rings, 40

Superconducting super collider, 74

Superconducting wire, 72

Superconductivity,

excitonic mechanism for, 84

Superconductivity mechanism, 227

effect of oxidation state, 114

effects of electro-negativity, 114

Superconductors: flux density in, 47

applications of, 73

critical fields of, 49

intermediate state in, 48, 55

mixed state in, 55

oxide, 1, 2

resistance of, 40

specific heat of, 40

synthetic methods, 69

thermal conductivity of, 40

thermopower of, 40

thin films of, 61

type I, 48

type II, 48, 86, 227

wires, 61

Superionic conductors, 115

Synthesis, 69

  of $YBa_2(Cu_{1-x}Ni_x)_3O_{7-\delta}$, 132

  of $YBa_2Cu_3O_{6.9}F_{0.11}$, 146

  of $YBa_2Cu_3O_{7-x}$, 131

  of $YBa_2Cu_3O_{7-z}$, 268

  of $YBa_2Cu_3O_xF_y$, 143

  of $YBa_cu_3O_{6.2}F_{0.63}$, 143

Synthetic methods:

  by chemical precipi-
    tation routes, 121

  carbonate method, 153

  citrate method, 153

  for oxyfluorides, 145

  for $YBa_2Cu_3O_{7-x}$, 121

  oxalate precipita-
    tion, 153

  precipitation methods,
    122

  sol-gel techniques, 153

Synthetic strategies, 108

T

T_c:

  Ag_7O_8X, 2

  $Ag_7O_8(HF_2)$, 10

  A_xMoO_3, 2

  A_xReO_3, 2

  A_xWO_3, 2

  $BaPbO_3$, 6

  $Ba(Pb,Bi)O_3$, 2, 6

  $BaPbO_{.75}Bi_{0.25}O_3$, 9

$BaPb_{1-x}Bi_xO_3$, 8

  effects of pressure
    on, 59

  effects in
    $R_{1+x}Ba_{2-x}Cu_3O_7$, 20

  of elements, 39

  $GdBa_2Cu_3O_x$, 330

  $HoBa_2Cu_3O_{6+x}$, 69

  K_xMoO_3, 5

  K_xMoO_3, under pres-
    sure, 5

  $K_{0.3}ReO_3$ 5

  $La_{1.25}Ba_{1.75}Cu_3O_{7.1}$, 20

  $La_{1.85}Ca_{0.15}CuO_4$, 60

  $(La,A)_2Cu_2O_4$, 2

  $(La,R,Sr)_2CuO_4$, 30

  $Li_{1+x}Ti_{2-x}O_4$ 5

  $LiTi_2O_4$, 2, 5

  Nb_3Ge, 38

  NbO, 2

  $RBa_2Cu_3O_7$, 2

  $SrTiO_{3-x}$, 2, 3

  TiO, 2

  $YBa_2(Cu_{1-x}Ni_x)_3O_{7-\delta}$, 135

  $YBa_2Cu_3O_{7-y}$, 202

Ternary copper oxides, 111

Ternary oxide system, phase
  diagram for, 352

Tetragonal vs. orthorhombic
  symmetry, 17, 19, 20,
    182, 211, 218

Tetragonal-to-orthorhombic
  transition: 11, 12, 19

  in $La_{2-x}Sr_xCuO_4$, 216,

and meanfield theory, 221

in $Ba_2Cu_3O_xF_y$, 143, 138

Tetramethylammonium carbonate, use in synthesis, 124

Tetramethylammonium hydroxide, 124

Thermal amplitudes of motion, 111, 112

Thermal conductivity, temperature dependence of, 43

Thermogravimetric analysis, of $YBa_2Cu_3O_{7-z}$, 269

Thermogravimetry, of $YBa_2Cu_3O_{7-z}$, 267

Thermopower, temperature dependence of, 43

Thick films, 159
    circuits, 160
    technology, advantages of, 160
    of $YBa_2Cu_3O_{6+x}$, 162

Thin films, superconducting, 61

TiO, 1
    metal-metal distance in, 3

$TmBa_2Cu_3O_7$, infrared reflectance of, 283

Trains, levitated, 74

Transmission lines, superconductive, 74

Transmission of electrical power, 74

Tungsten bronzes: 4
    cubic, 4
    cubic, 5
    hexagonal, 4
    stoichiometry of, 4
    superconducting, 4
    tetragonal, 4

Tunneling, 54, 89

Twin boundaries, 18, 19

Twins, 345

Two-point resistance, 254

Type I vs. Type II super-conductors, 49

U

Unit cell parameters:
    for $La_{1.85}Sr_{0.15}CuO_4$, 219
    for $YBa_2Cu_3O7_{0-x}$, 144

Unit cell volume, of $YBa_2(Cu_{1-x}Ni_x)O_{2-o}$, 138

V

$V_2O_{4-x}F_x$, 148

Valence band, O 2p, 6

Van der Pauw technique, 284

Vibrational excitations, 212

Vibrational spectrum, 223

Vibronic interaction, 105

Virtual magnetic excita-tions, 85

Voltage-current charac-teristics, 253

Vortex lines, 57

$W_{1-x}Re_xO_3$, 5

W

Weak link, 239

Wells' Salt, 229

Wetting angles, 341

Wires, superconducting, 61

Wolfram's Red Salt, 229

Work function barrier, 100

X

X-ray absorption, 169

X-Ray diffraction, of
$YBa_2Cu_3O_{7-z}$, 267

X-ray powder diffraction
data:
for $GdBa_2Cu_3O_{7-z}$, 329
for $YBa_2Cu_3O_{6.2}$, 144
for $YBa_2Cu_3O_{6.2}F_{0.63}$,
144
for $Y_1Ba_2Cu_3O_{7-z}$, 270

XANES, 168 ff.
of $La_{2-x}Sr_xCuO_4$, 174
of $YBa_2Cu_3O_4$, 174
of $YBa_2Cu_3O_{6.31}$, 174
of $YBa_2Cu_3O_{6.46}$, 174
of $YBa_2Cu_3O_{6.84}$, 174

XPS analysis, 255

XRD spectrum:
of $YBa_2Cu_3O_{6+x}$ film, 162
of $YBa_2Cu_3O_{6+x}$ film, 164

Y

$Y_{1.2}Ba_{0.8}CuO_4$, 68

$Y_1Ba_2Cu_3O_{6.85}$, 270

$Y_1Ba_2Cu_3O_{7-x}$, 236

$Y_1Ba_2Cu_3O_{7-x}$, current-
voltage characteris-
tics, 237

$Y_1Ba_2Cu_3O_{7-x}$, 240

$Y_2Ba_5Cu_7O_x$, 357

Y_2BaCuO_5, 338

Y_2BaCuO_5, 349
Raman spectrum of, 362

$Y_2Cu_2O_5$, 350, 351, 357
Raman spectrum of, 358

Y_2O_3, Raman spectrum
of, 355

Y_2O_3-BaO-CuO system, 357
phase diagram for, 352

$YBa_2(CU_{0.95}Ni_{0.05})_3O_{7-\delta}$,
critical field, 137, 138

$YBa_2(Cu_{1-x}Ni_x)_3O_{7-\delta}$:
grain size effects
in, 136
Meissner effect in, 136
oxygen stoichiometry
in, 136
resistivity of, 135
synthesis of, 132
T_c, 135
X-ray powder patterns
for, 133

$YBa_2Cu_{2.91}O_{7-\sigma}$, microstructures of, 342

$YBa_2Cu_3O_4$, XANES data
for, 174

$YBa_2Cu_3O_{5.6}$, 323

$YBa_2Cu_3O_6$, structure of, 16

$YBa_2Cu_3O_{6.2}$, X-ray powder
  diffraction data, 144
$YBa_2Cu_3O_{6.2}F_{0.25}$, 145
$YBa_2Cu_3O_{6.2}F_{0.54}$, 145
$YBa_2Cu_3O_{6.2}F_{0.63}$, 145
  X-ray powder diffrac-
    tion data, 144
$YBa_2Cu_3O_{6.31}$, XANES data
  for, 174
$Yba_2Cu_3O_{6.4}$, 243
  conductance of, 246, 247
  resistance of, 244, 245
$YBa_2Cu_3O_{6.46}$, XANES data
  for, 174
$YBa_2Cu_3O_{6.6}$, 202
$YBa_2Cu_3O_{6.84}$, XANES data
  for, 174
$YBa_2Cu_3O_{6.9}$, 243
  resistance of, 244
$YBa_2Cu_3O_{6.9}F_{0.11}$, 145
  synthesis of, 146
$YBa_2Cu_3O_{6+\delta}$, 305,323
$YBa_2Cu_3O_{6+x}$, 252, 349
  crystal structure of,
    357
  Raman scattering of, 363
  thick film of, 162
  XRD spectrum of, 162,
    164
$YBa_2Cu_3O_7$, 184, 213, 262
  Cooper pairs in, 23
  formulation of, 110
  infrared reflectance of,
    283
  isotope effect in, 23

  structure of, 16
$YBa_2Cu_3O_{7-\sigma}$:
  densification data for,
    340
  melting point, 341
  microstructure of, 335
  processing of, 335
$YBa_2Cu_3O_{7-\delta}$, 89, 228, 229
  fluoride substitution
    in, 141
  lattice parameters, 133
  solubility of Ni in, 132
$YBa_2Cu_3O_{7-x}$, 305
  isotope effect in, 82
  structure of, 131
  synthesis of, 131
  thermogravimetric
    results for, 125
  unit cell parameters
    for, 144
$YBa_2Cu_3O_{7-y}$, 184, 186, 187,
    200
  CuO_2 layers in, 203
  neutron diffraction
    structure, 185
  oxygen atom vacancies
    in, 183
  structure of, 181, 182
$YBa_2Cu_3O_{7-z}$, 268, 328
  electron microscopy, 267
  EPR of, 267, 271
  preparation of, 268
  resistance of, 271
  thermogravimetric
    analysis of, 269

thermogravimetry of, 267

X-ray diffraction of,
267, 270

$YBa_2Cu_3O_{7-z}$, X-ray powder
diffractogram, 270

$YBa_2Cu_3O_x$, 167, 174

peroxide formation in,
21

preparation at 800C, 21

$YBa_2Cu_3O_xF_y$, synthesis of,
143

$YBa_2Cu_3O_y$, 60

$YBa_3Cu_2O_{7+x}$, 350

$YBa_3Cu_3O_7$, 68

$YBa_3Cu_4O_x$, 357

$YBa_cu_3O_6.2F_{0.63}$, synthesis
of, 143

$YBaCuO_5$, green 211 phase
of, 362

$YbBa_2Cu_3O_{6+\delta}$, specific heat
of, 321

$YbBa_2Cu_3O_{7-x}$:

magnetic anomalies in,
313

magnetic susceptibility
of, 313

magnetization of, 319